Luana Lucas de Sá Almeida Veloso
Reginaldo Gomes Nobre

Production of graviola seedlings under salt stress and nitrogen depletion

Luana Lucas de Sá Almeida Veloso
Reginaldo Gomes Nobre

Production of graviola seedlings under salt stress and nitrogen depletion

Study on growth, morphophysiology and quality of seedlings

Imprint
Any brand names and product names mentioned in this book are subject to trademark, brand or patent protection and are trademarks or registered trademarks of their respective holders. The use of brand names, product names, common names, trade names, product descriptions etc. even without a particular marking in this work is in no way to be construed to mean that such names may be regarded as unrestricted in respect of trademark and brand protection legislation and could thus be used by anyone.

Cover image: www.ingimage.com

This book is a translation from the original published under ISBN 978-613-9-64904-4.

Publisher:
Sciencia Scripts
is a trademark of
Dodo Books Indian Ocean Ltd. and OmniScriptum S.R.L publishing group

120 High Road, East Finchley, London, N2 9ED, United Kingdom
Str. Armeneasca 28/1, office 1, Chisinau MD-2012, Republic of Moldova, Europe
Printed at: see last page
ISBN: 978-620-7-84621-4

SUMMARY

SUMMARY

The cultivation of soursop (Anonna muricata L.) in the irrigated areas of north-eastern Brazil is of great socio-economic importance. However, due to the limitation of low electrical conductivity water for irrigation in this region, it is necessary to study techniques that make it feasible to use saline water to grow the species. The aim of this study was to assess the tolerance of soursop seedlings to an increase in the salinity of irrigation water under fertilisation with different doses of nitrogen. The experiment was carried out in a greenhouse at the Federal University of Campina Grande, at the Centre for Agri-Food Sciences and Technology, Pombal campus, PB. A randomised block experimental design was used, in a 5 x 4 factorial scheme, with four replications and two plants per plot. The treatments referred to five levels of electrical conductivity of the irrigation water (0.3; 1.1; 1.9; 2.7 and 3.5 dS m^{-1}) in combination with four doses of nitrogen (70, 100, 130 and 160% of N of the recommended dose for soursop seedlings), in which the Morada Nova cultivar was evaluated. The following variables were assessed at 45 and 90 days after treatment (DAT): plant height, stem diameter, number of leaves, leaf area, leaf area ratio, specific leaf area and the physiological variables relating to the absolute and relative growth rate of stem diameter in the 15-45 and 45-90 (DAT) periods. In turn, the variables of fresh and dry phytomass of stem and leaves, dry phytomass of root, aerial part and total, as well as the Dickson quality index were evaluated at 90 (DAT). The growth of Morada Nova soursop seedlings subjected to different levels of water salinity was less affected in the initial phase (45 DAT). In the production of soursop seedlings, water with an ECa of 2.1 dS m-^{1} can be used, as it provides an acceptable average 10% reduction in growth. Doses of N higher than 70 mg of N dm^{-3} of soil do not attenuate salt stress or promote greater growth in Morada Nova soursop seedlings. The interaction between nitrogen doses and water salinity levels did not affect the production phase of Morada Nova soursop seedlings.

Keywords: Annona muricata L., saline water and fertiliser management.

CHAPTER 1

1 GENERAL INTRODUCTION

Fruit growing is an activity of great importance to the Brazilian agricultural sector. In recent years, Brazil has emerged as a major producer of fruit, particularly tropical and subtropical fruit, and is among the world's top three fruit producers, with an annual production of over 40 million tonnes. It is estimated that the fruit production chain covers three million hectares and generates six million direct jobs, highlighting the importance of fruit growing in Brazilian agribusiness (SEAB/DERAL, 2013).

Among these fruit trees, the soursop (Annona muricata L.), which belongs to the anonaceae family, has great economic value, especially in the Northeast region of Brazil where it is widely grown due to the favourable soil and climate conditions for its cultivation, as well as being highly appreciated by the population thanks to its flavour, aroma and pharmaceutical characteristics (FREITAS et al., 2013).

It can be propagated sexually or asexually, but producers have favoured seminal propagation of the soursop due to the difficulty of propagating it by grafting and the fact that soursop of seminal origin bears fruit within two years of planting. In addition, in the Northeast most of the production is sold to agro-industries, so there is no requirement for uniformity of fruit in terms of weight and shape (FIGUEREDO et al., 2013).

The fact that the Northeast has favourable soil and climate conditions for the production of soursop seedlings is not enough to give this region great potential for agricultural exploitation, as this progress is limited by its rainfall patterns (SANTANA et al., 2003; SILVA et al., 2015), characterised by prolonged periods of drought and irregular annual rainfall, where there is a water deficit for plants because the rate of evapotranspiration exceeds that of precipitation for most of the year (HOLANDA et al. 2010), which causes saline levels to rise in water sources.As a result, irrigation using water with a high concentration of soluble salts, especially sodium, has become common practice. Furthermore, using this water for a prolonged period of time ends up causing negative effects on the soils and crops planted in these areas, jeopardising their growth and development due to the toxic, osmotic and nutritional imbalance effects (NEVES et al... 2009), 2009).

Furthermore, the formation of soursop seedlings in the semi-arid region of the Northeast can be optimised through the use of techniques that enable the management of soil and water degraded by salts. The most important of these is the increase in nitrogen fertilisation in the production of seedlings, as this nutrient not only positively affects the vegetative growth of plants, but also promotes greater tolerance of plants to drought and salinity (POMPEU JÚNIOR, 1991; SÃO JOSÉ et al., 2014).

These benefits can be explained by the influence of nitrogen on absorption capacity, synthesis of amino acids, proteins, nucleic acids and chlorophylls (TAIZ & ZEIGER, 2013). In addition, the accumulation of these organic solutes increases the plants' osmotic adjustment capacity to salinity and increases the crops' resistance to salt and water stress (SILVA et al., 2008).

2 OBJECTIVES

2.1 General objective

To evaluate the effect of different doses of nitrogen on the growth of soursop seedlings irrigated with water of different levels of electrical conductivity - ECa.

2.2 Specific objectives

• To assess the effect of irrigation water salinity on the quality of seedlings fertilised with different doses of nitrogen.

• To determine the level of water salinity tolerated by the variables analysed in the soursop seedlings.

• To identify the best dose of nitrogen that favourably contributes to the growth of Morada Nova soursop seedlings.

3 LITERATURE REVIEW

3.1 Growing soursop (Annona muricata L.)

The soursop (Annona muricata L.) is a crop belonging to the genus Annona, native to the American continent, with its centre of origin in Central America and the Peruvian Valleys, and is considered the most tropical of the anonaceae (OKIGBO & OBIRE, 2009). It was spread by the Spanish and Portuguese from the Caribbean to south-eastern Mexico and Brazil (SILVA et al., 2013).

It is widespread in the tropical and subtropical regions of America, Europe, Asia and Africa (SACRAMENTO et al., 2009). It was introduced to Brazil in the 16th century and has become one of the most economically important fruit trees in Brazil, especially in the North, Northeast, Midwest and Southeast regions, especially in the states of Bahia, Alagoas, Ceará, Paraíba, Pernambuco and Pará, where there are favourable soil and climate conditions for its cultivation (SÃO JOSÉ et al., 2014).

Data on the area planted and the commercialisation of graviola are not precisely known, however, it is known that the cultivation of this fruit has advanced a great deal in Brazil in recent years, especially in the state of Bahia, in the region of the municipality of Irecê, whose production of graviola is 8,000 tonnes, with prospects for growth due to the increase in planted areas in the state (ADAB, 2010).

The cultivation of soursop is of great socio-economic importance due to the generation of employment and labour fixation, as production is not only destined for the agro-industry, but also has a significant volume sold as fresh fruit, especially on the domestic market (SÃO JOSÉ et al., 2000).

Of food importance, it is a fruit tree that has stood out due to its sensory characteristics of flavour and aroma, widely used both for fresh consumption and for agro-industry to obtain pulp, juice, nectar, among others (OLIVEIRA NETO et al., 2014; NOBRE et al., 2003). The fruit is also rich in vitamin C, calcium, carbohydrates, water and substances with antioxidant activity, the latter of which has received a great deal of attention because it helps protect the human body against oxidative stress, avoiding and preventing a series of chronic-degenerative disorders (TACO, 2006; YAHIA, 2010).

The soursop has an erect growth habit, with an average height of 4 to 8 metres in the adult stage, a single stem and asymmetrical branching. The leaves have a short petiole, are oblong-lanceolate or elliptical, 14 to 16 cm long and 5 to 7 cm wide; the veins are barely perceptible (MANICA, 1997).

This anonaceae has the largest root system among the other fruit trees in the family and can adapt to different types of soils, although it requires deep, rich, well-drained soils with a slightly acidic pH (5.5 - 6.5) (RAMOS, 1992). The fruit varies in shape and can be ovoid, condiform or irregular, with a white pulp that has a strong, acrid odour when green, becoming soft, pleasant, juicy, cooling, sweet, slightly acidic and somewhat cottony when ripe (CASTRO et al., 1984).

The seeds have exogenous dormancy, caused by the hardness of their outer skin, requiring scarification and/or immersion in cold water for 24 hours for perfect germination. Planting spacing can be from 4.0 x 4.5 m to 8.0 x 8.0 m (RAMOS, 1992).

3.2 Quality of seedlings

One of the factors that contributes to the formation of a highly productive orchard depends largely on the quality of the seedlings that will form the orchard (LIMA et al., 2007). Thus, when determining which seedlings are considered ready for planting, the parameters used are based on either morphological or physiological aspects. The morphological and physiological quality of seedlings depends on the genetic load and origin of the seeds, environmental conditions and production methods (AZEVEDO, 2003).

Morphological characteristics are the most widely used criteria for determining quality standards, but they still need to be better defined in order to meet the requirements of seedling survival and growth, which are determined by the adversities encountered in the field after planting. Among these morphological characteristics, the most studied are: shoot height, neck diameter, root and shoot dry matter weight and total weight, as well as the relationship between these parameters and quality indices, which are relevant for better evaluating the seedling production process (GOMES et al., 2001; CARNEIRO et al. 2007).

With regard to grapevine seedlings in general, they are considered ready to be transplanted into the field when they reach a height of 30-40 cm, due to their higher survival rate in the field as a result of their greater resistance to adverse environmental conditions, such as veranicos, competition with weeds and pest attacks (BARBOSA et al., 2003).

In terms of diameter, seedlings with a larger diameter are more likely to survive, as they are more efficient at forming and growing new roots. In general, the appropriate neck diameter for seedlings of fruit species considered to be of good quality for grafting is around 4.5 mm (TOMAZ, 2013).

3.3 Irrigation water quality

When studying water quality, it is difficult to find a simple and complete definition due to the great complexity of the determining factors and the large number of variable options available to describe irrigation water conditions. Despite this, in order to correctly interpret water quality for agricultural purposes, the parameters analysed must be related to their effects on the soil, the crop and irrigation management, which will be necessary to control or compensate for problems related to water quality (MELO, 2005; BERNARDO et al., 2006).

When it comes to the quality of water for irrigation, salinity is usually taken into account, since this parameter is related to the salinisation and sodification of soils, causing a drop in crop yields and a loss of the soil's chemical, physical and biological properties, forming degraded areas that cannot be cultivated (ALMEIDA, 2010; RIBEIRO, 2010; CARMO et al., 2011).

According to Ayers & Westcot (1999), irrigation water is classified according to three parameters: the first refers to the risk of salinity, and occurs when there is an accumulation of salts in the root zone at a certain concentration, causing a drop in yield; the second parameter refers to the risk of sodicity or infiltration problems, which occurs when relatively high levels of sodium or low levels of calcium and magnesium in the soil and water, through the dispersing action of sodium in the soil colloids, reduce hydraulic conductivity; and the third concerns toxicity from specific ions such as sodium, chloride and boron, which accumulate in plants in high concentrations, reducing the yield of crops sensitive to these ions.

Regarding their chemical composition, water in nature, whatever its source, is made up in greater proportions of sodium, calcium, magnesium and potassium salts, in the form of chlorides, sulphates, bicarbonates and carbonates, although the quantity and type of these salts can vary greatly depending on the source, geographical location and time of collection (RICHARDS, 1954; MEDEIROS et al., 2003). Chemical indicators of water quality include pH (hydrogen potential), alkalinity, hardness, chlorides, iron and manganese, nitrogen and phosphorus, among others (FONTES et al., 2011).

According to Maas (1990), water quality is only one of the factors that determine the type and intensity of management practices for the safe use of saline water in irrigation; other factors such as crop tolerance to salinity, soil properties, climate and crop management must be considered.

In the north-east of Brazil there are a large number of wells whose water is used for irrigation, representing an important input in the production chain. However, the use of poor quality water can cause damage to crops and soils, with serious socio-economic repercussions. The quality of the water from these wells can vary over time and space, with saline levels being higher during the dry season, as well as higher temperatures and evapotranspiration in the region (LARAQUE, 1989; SILVA JÚNIOR et al., 1999).

A total of 85% of the areas in the north-east of the country are located on impermeable crystalline rocks, where groundwater with a high salt content is located in the fractures of the rocks. However, the quality of this water can vary significantly depending on the type and quantity of dissolved salts, including the slow dissolution of limestone, gypsum and other minerals, which are transported by irrigation water and deposited in the soil, where they accumulate as the water evaporates or is consumed by crops. Furthermore, the fact that the annual rainfall rate in the region is lower than the evapotranspiration rate ends up causing a sharp increase in the accumulation of salts in the region's water sources, in general, with the predominance of NaCl, CaCl2 and MgCl2, in a ratio of 7:2:1 (MEDEIROS, 1992; AYERS; WESTCOT, 1999; FIGUEIREDO et al., 2009).

The use of substandard water and inadequate irrigation and drainage management has caused damage both to soils and plants and to irrigation systems, especially localised ones. This demonstrates the importance of monitoring water for irrigation and emphasises that the use of poor quality water for agriculture represents a threat to sustainability and public health (GARCIA et al., 2008; GHUNMI et al., 2009).

3.4 Salinity and its effects on crops

The use of saline water in agriculture should be considered an important alternative to the use of good quality water, which is currently seen as a scarce natural resource. However, salts in irrigation water or those already in the soil can have harmful effects on plants in their different phases (CAVALCANTE et al., 2001).

In general, salt stress inhibits plant growth by reducing the osmotic potential of the soil solution, causing a reduction in the availability of water for plants. It can also cause ionic toxicity, nutritional imbalances or both, due to the excessive accumulation of NaCl ions in plant tissues (MUNNS, 2002; FLOWERS & FLOWERS, 2005).

The presence of a high level of soluble salts in the rhizosphere causes changes in the physiological responses of plants (PARIDA & DAS, 2005). As a result, plants close their stomata to reduce water loss through transpiration, resulting in a lower photosynthetic rate, which is one of the causes of the reduced growth of species under conditions of salt stress (O"LEARY, 1971). In this way, NaCl affects the synthesis and translocation to the aerial part of the plant of hormones synthesised in the roots, which are indispensable for leaf metabolism (PRISCO, 1980). However, tolerance to salinity varies between and even within species depending on various factors, such as phenological stage, intensity and duration of salt stress (NEVES et al., 2009, BUSTINGORRI & LAVADO, 2011).

During the seedling stage, graviololeira adjusted osmotically as a moderately salt-tolerant plant and its leaf area and the biological yield of the plants increased with the level of salinity of the irrigation water up to 3.0 dS m^{-1} (CAVALCANTE et al., 2001).

However, Sá et al. (2015), when assessing the salt balance and initial growth of pine seedlings under substrates irrigated with saline water, found that high salt concentrations in the irrigation water inhibited the emergence, growth and phytomass formation of pine plants.

Salinity levels of up to 5.5 dS m^{-1} in irrigation water do not affect the germination of Morada Nova soursop seeds, but levels above 2.5 dS m^{-1} significantly reduce their speed of emergence (NOBRE et al., 2003). This can be explained by the fact that the salts present in the irrigation water reduce the osmotic potential of the soil solution, resulting in a delay in the soaking time of the seeds.

The increase in the soil's electrical conductivity caused by the salinity of irrigation water negatively affects seed germination, growth and production of the yellow passion fruit, which has led to the need for studies that can minimise the deleterious effects of salinity and, in addition, partially or totally replace the saline supply from mineral fertilisation (CAVALCANTE et al. 2001, 2002 and 2005).

3.4.1 Osmotic effect

The effects caused to plants by the high concentration of soluble salts, both in the soil and in the irrigation water, mean that studies must immediately turn to the osmotic component, as it is of great importance for the plant's absorption of water (WILLADINO & CAMARA, 2010).

Increasing the salinity of the water used for irrigation will lead to a significant increase in the salt content in the soil's saturation extract, resulting in a decrease in the osmotic potential of the soil solution, which is identified as the first factor in reducing growth (MUNNS, 1993; FLOWERS, 2004).

According to Dias & Blanco (2010), plants remove water from the soil when the soaking forces of the root tissues are greater than the forces with which the water is retained in the soil. The presence of salts in the soil solution causes the retention forces to increase due to their osmotic effect, giving rise to water stress.

This will have a number of implications for the plant, such as the non-absorption of nutrients, a reduction in the photosynthetic rate, a reduction in cell expansion, net CO_2 assimilation and accelerated senescence of mature leaves, consequently reducing the area destined for the photosynthetic process and the total production of photoassimilates (MUNNS, 2002; LACERDA et al., 2003).

According to Munns (2002), the osmotic effect can establish a new, lower rate of leaf and root elongation in a matter of hours, causing, over time, changes in the start of flowering and a reduction in seed production. As a result, an osmotic adjustment in the plant cell is necessary to ensure that turgour is maintained and that water can enter for cell growth. One of the mechanisms commonly cited for salinity tolerance has been the ability of some plants to accumulate ions in the vacuole and/or low molecular weight organic solutes in the cytoplasm, which can allow water absorption and cell turgidity to be maintained (TESTER & DAVENPORT, 2003; TAIZ & ZEIGER, 2004).

This compartmentalisation of salt allows tolerant plants to live in saline environments, but salinity-sensitive plants tend to exclude salts in the soil solution, but are unable to make the osmotic adjustment described and suffer from a decrease in turgor, leading to water stress, osmosis and death (DIAS & BLANCO, 2010).

3.4.2 Toxic effect

When certain ions from the soil or water are absorbed by plants and accumulate in their tissues in concentrations high enough to cause damage to the crop and reduce its yield, these plants suffer from toxicity (SILVA, 2011). Generally, most crops have evolved under conditions of low soil salinity, so the mechanisms developed to absorb, transport and utilise the nutrients present in non-saline substrates may not be effective in saline conditions (GARCIA et al., 2007).

Toxicity in plants can be caused by certain elements, such as sodium, boron, bicarbonates and chlorides, which in high concentrations cause physiological disturbances (BATISTA et al., 2002). In these toxic conditions, the concentration of Na^+ and/or Cl^- ions often exceeds the concentrations of macro and micronutrients, reducing the absorption of these mineral nutrients, especially NO^{3-}, $K+$ and Ca^{2+} (LARCHER, 2000).

In saline environments, NaCl is generally the predominant salt and consequently the one that causes the most damage to plants. In view of this, excess Na^+ and, above all, Cl^- in the protoplasm cause disturbances in the ionic balance, in addition to the specific effects of these ions on enzymes and cell membranes (FLORES, 1990). Therefore, the deleterious effects caused by the toxicity in plants can be expressed physiologically, leading to morphological reflexes, since the high concentration of ions in the transpiratory flow causes damage to the leaves, as well as early senescence (MUNNS, 2005; SILVA et al., 2008).

In addition, the symptoms of specific ion toxicity in the leaves are reported by Dias & Blanco (2010): A) the chloride symptom is evidenced by the burning of the leaf apex, reaching the edges in more advanced stages, promoting premature fall; B) typical sodium symptoms appear in the form of burns or necrosis along the edges in older leaves, progressing in the interneval area to the centre of the leaf as it intensifies; C) the symptoms caused by boron on the leaf are summarised as yellow or dry spots on the edges and apex of the old leaves, extending through the interneural areas to the centre of the leaf.

3.4.3 Nutritional effect

In addition to the osmotic and toxic effects caused to plants by excess salts in the soil or irrigation water, it is necessary to highlight another aspect affected by salinity that consequently affects crop growth and development. Nutritional imbalance occurs due to the significant alteration in the processes of absorption, transport, assimilation and distribution of nutrients in the plant, for example, excess Na inhibits the absorption of nutrients such as K and Ca. In addition, the high pH found in saline soil reduces the availability of many micronutrients such as copper (Cu), iron (Fe), manganese (Mn) and zinc (Zn) (FARIAS et al., 2009).

When the plant is exposed to saline soil for a long time, it experiences specific ionic phytotoxicity symptoms due to the excess absorption of Na^+ or Cl^-, causing an ionic imbalance, interfering with the stomatal mechanism and causing disturbances in metabolic activities in general (MANSOUR; SALAMA, 2004).

In aubergine, salt stress caused an increase in the levels of Na^+ and Cl^- ions in the leaves, followed by a reduction in the levels of Ca^{2+}, Mg^{2+} and K^+, reflecting the nutritional imbalance caused by progressive salt stress, which also decreased the concentration of K^+ and increased the levels of N, Cu^{2+}, Na^+ and Cl^- in the stem (BOSCO et al., 2009).

3.5 Plant tolerance to salinity

In terms of their response to salt stress, plants can be classified into two groups: halophytes and glycophytes. Halophytes are those that grow naturally in environments with high salt concentrations (typically Na^+ and Cl^-) and glycophytes are those that are unable to grow in environments with high salt concentrations (ORCUTT & NILSEN, 2000).

Among species sensitive to salt stress, the effect of salinity is manifested by severe reductions in growth and disturbances in membrane permeability, water exchange activity, stomatal conductance, photosynthesis and ion balance (NAVARRO et al., 2003; CABANERO et al., 2004). This deleterious effect of excess salts in the soil and irrigation water can significantly reduce crop yields and its magnitude depends on time, ion concentration, water use by crops and plant tolerance (WILLIDIANO & CAMARA, 2010).

The survival capacity of plants sensitive to saline stress is governed by mechanisms that confer resistance to salinity. These mechanisms are fundamental aspects of crop growth and involve high metabolic activity under moderate stress and low activity under severe stress, which allows the plant to withstand the stress (WILLIDIANO & CAMARA, 2010).

Plant tolerance to salinity depends on their ability to control salt transport at five specific points: 1- Selectivity in the absorption process by root cells; 2 - Loading the xylem preferentially with K^+, rather than Na^+; 3 - Removal of salt from the xylem in the upper part of the roots, stem, petiole or leaf sheaths; 4 - Retranslocation of Na^+ and Cl- in the phloem, ensuring no translocation to tissues of the aerial part in the process of growth and; 5 - Excretion of salts through glands or vesicular hairs, present only in halophytes. Tolerance in glycophytes depends on the first three mechanisms, and these occur to varying degrees depending on the species and/or cultivar (MUNNS et al, 2002).

One of the strategies used by plants is the extrusion of Na^+ into the soil solution, removing the cation from the plant, and the expulsion of Na^+ from some tissues, especially the xylem, as a way of avoiding the accumulation of the cation in the leaf limb, minimising the deleterious effects of salinity on leaf metabolism, especially the photosynthetic process (MUNNS et al., 2002). There are also some species that have the capacity to accumulate ions in the vacuole and low molecular weight organic solutes in the cytoplasm, in order to lower their water potential to a level below that of the soil, which allows them to make an osmotic adjustment to this type of condition. In turn, other crops show tolerance due to differences in the acquisition, translocation, transfer or accumulation of Na and Cl ions (FARIAS, 2008).

Compatible solutes make up a small group of substances of a different chemical nature: amino acids (such as proline), quaternary ammonium compounds (glycine betaine, β-alanine betaine, proline betaine, choline-O-sulphate), tertiary sulphonium compounds (DMSP - dimethylsulphoniopropionate), polyols (or polyhydric alcohols such as pinitol and mannitol), soluble sugars (fructose, sucrose, trehalose, raffinose) or polymeric sugars (fructans), as well as polyamines (putrescine, spermidine and spermine). Some enzymes that eliminate free radicals and proteins that protect the formation or stability of other proteins should also be included (WILLIDINO & CÂMARA, 2010).

Among the factors studied, the nutritional status of plants is a factor that can be taken into account to characterise plant tolerance to salinity, since increases in the concentration of NaCl in the soil solution impair root absorption of nutrients, especially K and Ca, and interfere with their physiological functions. Therefore, the ability of plant genotypes to maintain high levels of K and Ca and low levels of Na within the tissue is one of the key mechanisms that contributes to expressing greater tolerance to salinity. In most cases, salinity-tolerant genotypes are able to maintain high K/Na ratios in their tissues (DIAS & BLANCO, 2010).

As the application of fertilisers increases the concentration of nutrients in the soil, some authors state that the application of fertilisers in greater quantities than the recommended amount would bring benefits in conditions of moderate salinity, as there would be greater absorption of nutrients, increasing the K/Na, Ca/Na and NO3/Cl ratios, as Ca, N and K are nutrients that can confer a certain degree of tolerance to salinity on crops (CUATERO & MUNOZ, 1999).

3.6 Nitrogen fertilisation

Nitrogen is an essential element for plants and its deficiency limits crop growth and productivity, as it is required at all stages of plant development. Its insufficiency is observed in almost all soils, and the criterion for identifying deficiency is the appearance of generalised chlorosis on the leaves, starting with the oldest leaves, which is related to the participation of N in the structure of the chlorophyll molecule (CARVALHO et al., 2003; SILVA et al., 2010).

In the soil, nitrogen is available in organic (amino acids, peptides and insoluble complex forms) and mineral forms, with organic forms predominating (SOUSA & FERNANDES, 2006). The mineral nitrogen in the soil is represented by the ionic forms ammonium (NH_4^+), nitrate (NO_3^-), and very rarely, nitrite (NO_2^-), with the ammoniacal and nitric forms being those readily absorbed by plants (FURTINI NETO et al., 2001).

The use of nitrogen through nitrogen fertilisation not only promotes growth and good crop development, but can also reduce the effects of salinity on plant species. The explanation may be related to the functions of this element in plants, since it plays a structural role, forming part of various organic compounds that are vital for the plant, such as amino acids, proteins, among others (ALVES et al., 2012; DIAS et al., 2012).

According to Barbosa et al. (2003), nitrogen is one of the most demanding macronutrients for soursop seedlings. They found that the absence of N, Ca and P, in that order, led to the most negative results, reducing the height and diameter of the plant stems, as well as reducing the dry matter of leaves, stems and roots.

Among the nitrogen sources used in adduction, urea stands out. This is the most widely used nitrogen fertiliser in Brazil, mainly due to its low cost per unit of nitrogen and high N concentration (45%), which reduces transport costs. It also has a high N concentration, is easy to find on the market and requires two reactions to reach the nitrate form (hydrolysis and nitrification). This last characteristic tends to delay the start of the leaching process, since ammonium ($NH4^+$) is retained by the negative electrical charges of the soil before being nitrified (ERNANI, 2003).

Even though there is a great need to fertilise soursop trees properly, there are few experimental studies on the subject, and most of the fertilisation suggestions found are based on other crops and not on results obtained in the field by studying the species (SILVA & SILVA, 1997).

4 BIBLIOGRAPHICAL REFERENCES

AZEVEDO, M. I. R. Quality of pink cedar (Cedrela fissilisVell.) and yellow ipê (Tabebuia serratifolia (Vahl) Nich.) seedlings produced in different substrates and tubes. 90f Dissertation (Master's Degree in Forest Science) - Federal University of Viçosa, Viçosa, 2003.

ADAB. Agricultural Defence Agency of the State of Bahia. 2010 Available at: <http://www.adab.ba.gov.br/modules/news/article.php?storyid=480>. Accessed on: 21/05/2016.

ALMEIDA, O. A. **Qualidade da água de irrigação** - Cruz das Almas: Embrapa Mandioca e Fruticultura, v. 227, 2010.

ALVES, A. N.; GHEYI, H. R.; UYEDA, C. A.; SOARES, F. A. L.; NOBRE, R. G.; CARDOSO, J. A. F. Use of saline water and nitrogen fertilisation in the cultivation of BRS-energia papaya. **Revista Brasileira de Agricultura Irrigada**, v.6, n. 2, p. 151-163, 2012.

AYERS, R. S.; WESTCOT, D. W. Water quality in agriculture. In: GHEYI, H.R.;
MEDEIROS, J. L.; DAMASCENO, F. A. V. (Trad.). **FAO Studies: Irrigation and Drainage**, 29 Revised, Campina Grande: Federal University of Paraíba. 153 p.1999.

BARBOSA, Z.; SOARES, I.; CRISÓSTOMO, L. A. Growth and nutrient absorption by soursop seedlings. **Revista Brasileira de Fruticultura,** v. 25, n. 3, p. 519-522. 2003.

BATISTA, M. J.; NOVAES, F.; SANTOS, D. G.; SUGUINO, H. H. **Drainage as Instruments for desalination and prevention of soil salinisation**. 2nd edition, revised and expanded. Brasília: CODEVASF, 216p. 2002.

BERNARDO, S.; SOARES, A. A.; MANTOVANI, E. C. **Manual de irrigação**. 8. ed. Viçosa: UFV, 2006, p. 625.

BOSCO, M. R. O. de; OLIVEIRA, A. B. de, HERNANDEZ, F. F. F.. Influence of salt stress on the mineral composition of aubergine. **Revista Ciência Agronômica,** v. 40, n. 2, p.157-164, 2009.
BUSTINGORRI, C.; LAVADO, R. S. Soybean growth under stable versus peak salinity. **Scientia Agricola,** v.68, p.102-108, 2011.

CABANERO, F.J., MARTINEZ, V., CARVAJAL, M. Does calcium determine water uptake under saline conditions in pepper plants, or is it water flux, which determines calcium uptake. **Plant Science,** v.166, p.443-450, 2004.

CARNEIRO, J. G. A.; BARROSO, D. G.; SOARES, L. M. S. Growth of bare root seedlings of Pinus taeda, L., under five spacings in the nursery and their performance in the field. **Revista Brasileira de Agrociência,** v. 13, p. 305 - 310. 2007.

CARMO, G. A; OLIVEIRA, F. R. A.; MEDEIROS, J. F.; OLIVEIRA, F. A.; CAMPOS, M. S.; FREITAS, D. C. Leaf contents, accumulation and partitioning of macronutrients in the pumpkin crop irrigated with saline water. **Revista Brasileira de Engenharia Agrícola e Ambiental,** v.15, n.5, p.512-518, 2011.

CARVALHO , M. A. C.; FURLANI JUNIOR, E. O. A. Doses and times of nitrogen application and leaf contents of this nutrient and chlorophyll in bean. **Revista Brasileira de Ciências do Solo,** v.27, p.445-450, 2003.

CASTRO, F. A. de, MAIA, G. A.; HOLANDA, L. F. F. Physical and chemical characteristics of graviola, **Pesquisas Agropecuária Brasileira,** v.19, n.3, p. 361-365, 1984.

CAVALCANTE, L. F.; CARVALHO, S. S. DE , LIMA. E. M. DE. Initial development of soursop under sources and levels of water salinity. **Revista Brasileira Fruticultura,** , v.23, n.2, p. 455-459, 2001.

CAVALCANTE, L. F.; CAVALCANTE, I. H. L.; PEREIRA, K S. N.; OLIVEIRA, F. A. de; GONDIM, S C.; ARAÚJO, F A. R. de. Germination and initial growth of guava plants irrigated with saline water. R. de. Germination and initial growth of guava plants irrigated with saline water. **Revista Brasileira de Engenharia Agrícola e Ambiental**, v.9, n.4, p.515-519, 2005.

CAVALCANTE, L. F; VIEIRA, M. S.; SANTOS, A. F.; OLIVEIRA, W. M.; NASCIMENTO, J. A. M. Saline water and liquid bovine manure in the formation of Paluma guava seedlings. **Revista Brasileira de Fruticultura**, v.32, n.1, p.251-261, 2010.

CAVALCANTE, L. F.; SANTOS, J. B.; SANTOS, C. J. O.; FILHO, J. C. F.; LIMA, E. M. & CAVALCANTE, I. H. L. Seed germination and initial growth of passion fruit trees with saline water in different substrate volumes. **Revista Brasileira de Fruticultura**, v. 24, n. 3, p. 748-751, 2002.

CUARTERO, J.; MUNOZ, R. F. Tomato and salinity. **Scientia Horticulturae**, v.78, n.1/4, p.83-125, 1999.

DIAS, M. J. T.; SOUZA, H. A.; NATALE, W.; MODESTO, V. C.; ROZANE, D. E . Nitrogen and potassium fertilisation of guava seedlings in a commercial nursery. **Ciências Agrárias**, v.33, n.1, p. 2837-2848, 2012.

DIAS, N. S.; BLANCO, F. F. Effect of salts on soil and plants. In: GHEYI, H. R.; DIAS, N. S.; LACERDA, C. F. **Management of salinity in agriculture**: Basic and applied studies. Fortaleza, INCT Sal, 2010, p.129-14.

ERNANI, P. R. **Nitrogen availability and nitrogen fertilisation for apple trees.** UDESC, 2003.

FARIAS, S. G. G. **Osmotic stress on the germination, growth and mineral nutrition of glycyrrhiza *(Gliricidia sepium (Jacq.)*** 61f (Master's thesis) Universidade Federal de Campina Grande. Brazil, 2008.

FARIAS, S. G. G. D., SANTOS, D. R. D. U., & FREIRE, A. L. D. O. U. Salt stress on initial growth and mineral nutrition of Gliricidia (Gliricidia sepium (Jacq.) Kunt ex Steud) in nutrient solution. **Revista Brasileira de Ciência do Solo**, v.33, n.5, p.1499-1505, 2009.

FIGUEIRÊDO, G. R. G. D.; VILASBOAS, F. S.; OLIVEIRA, S. J. R. D.; SODRÉ, G. A.; SACRAMENTO, C. K. D. Propagation of soursop tree through minicuttings. **Revista Brasileira de Fruticultura**, v.35, n.3, p. 860-865, 2013.

FIGUEIREDO, V. B.; MEDEIROS, J. F.; ZOCOLER, J. L.; SOBRINHO, J. S. Watermelon crop evapotranspiration irrigated with water of different salinities. **Engenharia Agrícola,** v. 29, n. 02, p. 231-240, 2009.

FLORES, H. E. Polyamines and plant stress. In: LASCHER, R. G.; CUMMING, J. R. **Stress responses in plants: adaptation and acclimation mechanisms.** New York, Wileyliss, 1990, p. 217-239.

FLOWERS, T. J. Improving crop salt tolerance. **Journal of Experimental Botany,** v.55, p. 307-319, 2004.

FLOWERS, T. J.; FLOWERS, S. A. Why does salinity pose such a difficult problem for plant breeders? **Agricultural Water Management,** v. 78, n. 1, p. 15-24, 2005.

FONTES, R. L. F. F.; CANTARUTTI, R. B.; NEVES, J. C. L. **Fertilidade de solo.** Brazilian Society of Soil Science, 2011.

FREITAS, A. L. G. E.; VILASBOAS, F. S.; PIRES, M. M.; SÃO JOSÉ, A. R. Characterisation of Graviola *(Annona muricata* L.) Production and Market in the State of Bahia. **Economic Information,** v. 43, n. 3, p.23-34, 2013.

FURTINI NETO, A. E.; VALE, F. R.; RESENDE, A. V.; GUILHERME, L. R. G.; GUEDES, G. A. A. **Fertilidade do solo.** Lavras: UFLA/FAEPE, 2001, p. 261.

GARCIA, G. de O.; MARTINS FILHO, S.; REIS, E. F. dos MORAES, W. B.; NAZÁRIO, A. DE A. Chemical changes in two soils irrigated with saline water. **Revista Ciência Agronômica,** v.39, n.1, p. 7-18, 2008.

GARCIA, G. O. de; FERREIRA, P. A.; MIRANDA, G. V. Foliar contents of cationic macronutrients and their relationship with sodium in maize plants under salt stress. **IDESIA.** v. 25, n.3, p. 93-106, 2007.

GHUNMI, L. A.; ZEEMAN, G.; AYYAD, M.; LIER, V.J.B. Grey water treatment in a series anaerobic - aerobic system for irrigation. **Bioresourse Technology,** v. 101, n.1, p. 41-50, 2009.

HERNÁNDEZ, L. V.; IDAL, N. A. M.; MOCTEZUMA, H. L. Propuesta de un plan de desarrollo integral del guanábano (Annona muricata L.) en el estado de veracruz méxico. **Revista Brasileira de Fruticultura,** v. 36, n.1, p. 94-101, 2014.

HOLANDA FILHO, R. S. F.; SANTOS, D. B.; AZEVEDO, C. A. V. de.; COELHO, E. F.; LIMA, V. L. A.

de. Saline water on soil chemical attributes and nutritional status of mandioqueira. **Revista Brasileira de Engenharia Agrícola e Ambiental,** v.15, n.1, p.60-66, 2011.

HOLANDA, J. S.; AMORIM, J. R. A.; FRRREIRA NETO, M.; HOLANDA, A. C. Water quality for irrigation. In: GHEYI, H. R.; DIAS, N. S.; LACERDA, C. F (ed). **Salinity management in agriculture:** Basic and applied studies. Fortaleza, INCTA Sal, p. 472, 2010.

LACERDA, C.F.; CAMBRAIA, J.; CANO, M.A.O.; RUIZ, H. A.; PRISCO, J.T. Solute accumulation and distribution during shoot and leaf development in two sorghum genotypes under salt stress. **Environmental and Experimental Botany,** v.49, n.2, p.107, 2003.

LARAQUE, A. **Estudo e previsão da quantidade de água de açudes do Nordeste semi-árido brasileiro.** Recife: SUDENE, Hydrological Series, v.26, p.95, 1989.

LARCHER, W. **Ecofisiologia vegetal,** Ed. RiMa Artes e Textos, São Carlos, 2000, p.531.

LIMA, J. F.; PEIXOTO, C. P.; LEDO, C. A. S. Physiological indices and initial growth of papaya (Carica papaya L.) in a greenhouse. **Ciência e Agrotecnologia,** v. 31, n. 5, p.1358-1363, 2007.

MANICA, I. Taxonomy, morphology and anatomy. In: SÃO JOSÉ, A. R. et al. (eds.). **Anonáceas, produção e mercado (pinha, graviola, atemóia and cherimólia).** Vitória da Conquista: UESB. p. 20- 3, 1997.

MANSOUR, M.M.F.; SALAMA, K.H.A. Cellular basis of salinity tolerance in plants.**Environmental and Experimental Botany,** Elmsford, v.52, n.2, p.113-122, 2004.

MAAS, E. V. Crop salt tolerance. In: TANJI, K. K. **Agricultural salinity assessment and management.** New York: ASCE, 1990. ch. 13, p. 262-304.

MEDEIROS, J. F. **Quality of irrigation water and evolution of salinity in properties assisted by "GAT" in the states of RN, PB and CE.** Dissertation (Master's Degree in Agricultural Engineering) - Postgraduate Programme, Federal University of Paraíba, Campina Grande, 1992, p. 173.

MEDEIROS, J. F. DE; LISBOA, R. A.; OLIVEIRA, M.; SILVA JÚNIOR, M. J.; ALVES, L. P. Characterisation of groundwater used for irrigation in the melon-producing area of Chapada do Apodi. **Revista Brasileira Engenharia Agrícola e Ambiental,** Campina Grande, v. 7, n.3, p. 469- 472, 2003.

MELO, A. D. Operation of reservoirs in the semi-arid region considering water quality criteria. 2005. 87f. Dissertation (Master's in Civil and Environmental Engineering) - Federal University of Campina Grande, Campina Grande, 2005.

MUNNS, R.; SHARP, R. E. Involvement of abscisic acid in controlling plant growth in soils of low water potential. **Australian Journal of Plant Physiology**, v. 20, p. 425437, 1993.

MUNNS, R. Comparative physiology of salt and water stress. **Plant, and Cell Environment,** v.25, n.2, p.239-50, 2002.

MUNNS, R. Genes and salt tolerance: bringing them together. **New Phytologist**, v.167, p. 645-663, 2005.

NAVARRO, J.M., GARRIDO, C., MARTINEZ, V., CARVAJAL, M. Water relations and xylem transport of nutrients in pepper plants grown under two different salt stress regimes. **Plant Growth Regulators**, v.41, p.237-245, 2003.

NEVES, A. L. R.; LACERDA, C. F.; GUIMARÃES, F. V. A.; HERNANDEZ, F. F. F.; SILVA, F. B.; PRISCO, J. T.; GHEYI, H. R. Biomass accumulation and nutrient extraction by string bean plants irrigated with saline water at different stages of development. **Ciência Rural**, v.39, n.3, p. 758-765, 2009.

NOBRE, R. G.; FERNANDES, P. D.; GHEYI, H. R. SANTOS, F. J. D. S.; BEZERRA, I. L.; GURGEL, M. T. Germination and formation of grafted soursop seedlings under salt stress, **Pesquisa agropecuaria brasileira,** v. 38, n. 12, p. 1365-1371, 2003.

OKIGBO, R. N.; OBIRE, O. Mycoflora and production of wine from fruits of soursop (Annona Muricata L.). **International Journal of Wine Research**, v.1, p.1-9, 2009.

O'LEARY, J. W. High humidity overcomes lethal levels of salinity in hydroponically grown salt-sensitive plants. **Plant and Soil**, v.42, p.717-721, 1971.

OLIVEIRA NETO, E. A. de.; SANTOS, D. C. da.; SANTOS, Y. M. G. dos, Agroindustrial utilisation of soursop (Annona muricata L.) for production of liqueurs: Sensory evaluation, **Journal of Biotechnology and Biodiversity.** v. 5, n.1, p. 33-42, 2014.

ORCUTT, D. M.; NILSEN, E. T. **Physiology of Plants Under Stress**. New York, John Willey & Sons, 2000.

PARIDA, A. K.; DAS, A. B. Salt tolerance and salinity effects on plants: A review. Ecotoxicology and Environmental Safety, v.60, p.324-349, 2005.

POMPEU JÚNIOR, J. **Rootstocks**. In: RODRIGUES, O.; VIEGAS, F. C. P.; POMPEU JÚNIOR, J. Citricultura brasileira. 2. ed. Campinas: Fundação Gargill,. v.1, p.265-280, 1991.

PRISCO, J. T. Some aspects of the physiology of salt stress. **Revista Brasileira de Botânica,** , v.3, p.85-94, 1980.

RAMOS, V. H. V. Cultivation of the soursop tree (Annona muricata L.) In: DONADIO, L. C. Fruticultura tropical. Jaboticabal, FUNEP, 1992, p. 268.

RIBEIRO, M. R. Origin and classification of soils affected by salts. In: GHEYI, H. R.; DIAS, N. S.; LACERDA, C. F (ed). **Salinity management in agriculture: Basic and applied studies**. Fortaleza, INCT Sal, 2010, p. 472.

RICHARDS, L. A. (ed.) **Diagnosis and improvement of saline and alkali soils**. Washington, United States Salinity Laboratory: 1954. 160p. (USDA. Agriculture Handbook, 60).

SACRAMENTO, C. K.; MOURA, J. I. L.; COELHO JUNIOR, E. Graviola. In: SANTOS-SEREJO, J. A. et al. (eds.). **Tropical fruit growing: regional and exotic species**. Brasília, DF: Embrapa Informação Tecnológica, p. 95-132, 2009.

SANTANA, M. J.; CARVALHO, J. A.; SILVA, E. L.; MIGUEL, D. S. Effect of irrigation with saline water on a soil cultivated with bean (phaseolus vulgaris L.). **Ciência e agrotecnologia**, v.27, n.2, p.443-450, 2003.

SÃO JOSÉ, A. R.; PIRES, M.; FREITAS, A.; RIBEIRO, D. P.; PEREZ, L. A. A. Current events and prospects for Anonaceae in the world. **Revista Brasileira de Fruticultura**, v. 36, n. especial, p. 86-93, 2014.

SÃO JOSÉ, A. R.; PRADO, N. B. DO; BOMFIM, M. P. Nutrient uptake in anonaceae. **Revista Brasileira de Fruticultura**, v.36, p. 176-183, 2014.

SÃO-JOSÉ, A. R.; ANGEL, D. N.; BONFIM, M. P.; REBOUÇAS, T. N. H. Cultivation of graviola. In: Semana Internacional de Fruticultura e Agroindústria, v.7, 2000, Fortaleza. Courses. Fortaleza: Sindifruta, Instituto Frutal, p. 35, CD-ROM. 2000.

SEAB/DERAL. Fruit growing - Analysing the agricultural situation. Available at: http://www.agricultura.pr.gov.br/arquivos/File/deral/Prognosticos/fruticultura_2012_13. pdf. 25 Apr. 2013.

SILVA JUNIOR, F. C. **Manual de analise química de solos, plantas e fertilizantes.** Brasília: Embrapa Communication for Technology Transfer, p.370, 1999.

SILVA SÁ, F. V. da, BRITO, M. E. B., PEREIRA, I. B., NETO, P. A., ANDRADE SILVA, L. de, COSTA, F. B. da. Salt balance and initial growth of pine seedlings *(Annona squamosa* L.) under substrates irrigated with saline water. **Irriga,** v. 20, n.3, p. 544, 2015.

SILVA, A. Q.; SILVA, H. Current situation and prospects for anonaceae in the State of Bahia. In: SÃO JOSÉ, A. R. et al. (eds.). **Anonáceas:** Tecnologia de produção e comercialização, Vitória da Conquista, BA: DFZ/UESB, p.168-172. 1997.

SILVA, E. M. F.; NASCIMENTO, R. B. C. de; BARRETO, F. S. In vitro study of the cytotoxic potential of Annona muricata L. **Revista Ciências Farmaceitica Básica Aplicada,** v. 36, n.2, p.277-283, 2015.

SILVA, E. C.; NOGUEIRA, R. J. M. C.; ARAÚJO, F. P.; MELO N. F.; AZEVEDO NETO. Phisiologi-Phisiological responses to salt stress in Young umbu plants. **Enviromental and Experimental botany,** v. 63, p. 147-157, 2008.

SILVA, G. B. P. da ; LIMA, K. D. R. de; PROCÓPIO , I. J. S. Production of pine seedlings (Annona squamosa L.) under doses of ammonium sulphate. **Revista Verde,** v.5, n.5, p. 204-209, 2010.

SILVA, I. N. Water quality in irrigation. **ACSA - Agricultura Científica no Semi Árido,** v.07, n.3, p. 01-15, 2011.

SILVA, R. A. R. da.; NUNES, J. C.; LIMA NETO, A. J. de,. Irrigation rates and soil cover in the production and quality of soursop fruit. **Revista Brasileira Ciências Agrárias,** v.8, n.3, p.441-447, 2013.
SOUSA, S. R.; FERNANDES, M. S. Nitrogen. In: FERNANDES, M. S. **Nutrição mineral de plantas.** Viçosa: SBCS, p. 432, 2006.

TACO - Brazilian Food Composition Table. NEPA - UNICAMP - version II - 2ed - Campinas, SP : NEPA-UNICAMP, 2006.

TAIZ, L.; ZEIGER, E. **Plant physiology**. 3. ed.Porto Alegre: Artmed, Ed. 5. p. 918, 2013.

TAIZ, L.; ZEIGER, **Fisiologia vegetal,** ed. 3Porto Alegre, Artmed., p. 719, 2004.

TESTER, M.; DAVENPORT, R. Na+ tolerance and Na+ transport in higher plants. **Annals of Botany**, v. 91, p. 503-527, 2003.

TOMAZ, Z. F. P. Pelotas. 2013. 159f. **Cloning rootstocks and producing peach tree seedlings in soilless cultivation systems**. Thesis (Doctorate in Agronomy, area of concentration Temperate Climate Fruit Growing) - Eliseu Maciel School of Agronomy, Federal University of Pelotas.

WILLADINO, L.; CAMARA, T. R.L; Plant Tolerance to Salinity: Physiological and Biochemical Aspects. **Enciclopédia Biosfera, Centro Científico Conhecer,** v.6, n.11; p.21, 2010.

YAHIA, E. M. The contribution of fruit and vegetable consumption to human health. In L. A. Rosa, E. Alvarez-Parrilla, & G. A. Gonzalez-Aguilara (Eds.), **Fruit and vegetable phytochemicals chemistry nutritional value and stability.** Wiley-Blackwell: Hoboken, 2010.

CHAPTER 2

PRODUCTION OF SOURSOP SEEDLINGS IRRIGATED WITH WATER OF DIFFERENT SALINITIES AND NITROGEN FERTILISATION

SUMMARY

The soil and climate characteristics of the semi-arid region of north-eastern Brazil favour the exploitation of soursop. However, due to the limited quantity and quality of water for irrigation in this region, it is necessary to study techniques that make it feasible to use saline water to grow the species. The aim of this study was to assess the tolerance of soursop seedlings to salinity in irrigation water, under fertilisation with different doses of nitrogen. The experiment was carried out in a protected environment at the Federal University of Campina Grande, at the Centre for Agri-Food Sciences and Technology, Pombal-PB campus, in a randomised block experimental design, in a 5 x 4 factorial scheme, with the treatments referring to five levels of electrical conductivity of the irrigation water - ECa (0.3; 1.1; 1.9; 2.7 and 3.5 dS m^{-1}) in interaction with four doses of nitrogen - N (70, 100, 130 and 160% of N of the recommended dose). Four replications were used, with two plants per plot, where the "Morada Nova" cultivar was evaluated. The effect of the different treatments on morphological variables was assessed at 45 and 90 days after the treatments were applied (DAT). The growth of soursop seedlings subjected to different levels of water salinity was less affected in the initial phase (45 DAT). In the production of soursop seedlings, water with an ECa of 2.1 dS m-^{1} can be used, as it provides an acceptable average 10% reduction in growth. Doses of N higher than 70 mg of N dm"^{3} of soil do not attenuate salt stress, nor do they promote greater growth in Morada Nova soursop seedlings. The interaction between nitrogen doses and water salinity levels did not affect the growth phase of soursop seedlings.

Keywords: Annona muricata L., salt stress, mineral fertilisation.

1 INTRODUCTION

The Northeast of Brazil has favourable soil and climate conditions for the cultivation of various fruit species, including the soursop (Annona muricata L.), which is one of the most important of the Annonaceae family in terms of economic expression in the region due to its consumer preference (COSTA et al., 2008; SANTOS et al., 2014).

However, the Northeast region is characterised by low rainfall and high evaporation rates, naturally causing a water deficit and an increase in the saline concentration of water sources. In this sense, the use of water with a high concentration of salts has become frequent, compromising soil quality and crop yields

(NEVES et al., 2009). Furthermore, combined with the climatic factors of the northeast region, soil salinity has increased due to inadequate irrigation and fertiliser management (HOLANDA FILHO et al., 2011).

The increase in the concentration of salts in the soil above the value tolerated by most species has negative effects that can be observed throughout the crop stand, which can affect plants due to the reduction in the osmotic potential of the soil solution, ionic toxicity and nutritional imbalance, damaging physiological processes such as CO_2 assimilation, protein synthesis and, in extreme cases, plant death, which consequently limits productive capacity, resulting in serious damage to agricultural activity (SOUSA et al., 2011).

In this sense, knowledge of the average salt content in the root zone tolerated by plants at different stages of development can favour the use of water with a certain degree of salinity, which is so common in the northeast of Brazil. It is therefore essential to carry out studies aimed at obtaining crop tolerance indices to salinity, as well as discovering techniques to mitigate the harmful effects caused to crops by the high concentration of salts in irrigation water and/or soil (CAVALCANTE et al., 2010).

Among the techniques, nitrogen fertilisation has been used to increase plant tolerance to salinity, due to the nutrient's role in the production of amino acids, proteins, nucleic acids and chlorophylls, which favours this tolerance (LIMA et al., 2014). In addition, the accumulation of these organic solutes inside the cell can increase the plants' osmotic adjustment capacity, helping and favouring increased tolerance to certain levels of water and salt stress (NASCIMENTO et al., 2015), producing better quality seedlings for planting the orchard, increasing the rate of setting in the field, a fact of great importance, especially in the first few months after planting, when they are subjected to more adverse environmental conditions (OLIVEIRA et al., 2014).

The aim of this study was to evaluate the production of soursop seedlings irrigated with salinised water and fertilised with different doses of nitrogen.

2 MATERIAL AND METHODS

2.1 Location of the experiment

The work was carried out from December 2015 to April 2016 in a protected environment (greenhouse) at the Centre for Agri-Food Sciences and Technology of the Federal University of Campina Grande (CCTA/UFCG), Pombal Campus - PB, whose local geographical coordinates are 6°48'16" S, 37°49'15" O and an average altitude of 184 m.

According to the Koppen classification adapted to Brazil, the region's climate is of the hot semi-arid BSh type, with an average annual temperature of 28°C, rainfall of around 750 mm per year^{-1} and average annual evaporation of 2000 mm (COELHO & SONCIN, 1982).

2.2 Treatments and statistical design

The treatments consisted of five levels of electrical conductivity of the irrigation water - ECa (0.3; 1.1; 1.9; 2.7 and 3.5 dS m^{-1}) associated with four doses of nitrogen: (70, 100, 130 and 160% of N, which corresponds to 70, 100, 130 and 160 mg dm^{-3} of the dose recommended by Novais et al. (1991)). The treatments were distributed in a randomised block design, in a 5 x 4 factorial scheme, with four replications, each plot consisting of two plants.

2.3 Description of treatments

The salt levels were selected on the basis of a study by Cavalcante et al. (2001), who classified soursop in the initial growth phase as moderately sensitive to salinity, i.e. the biological yield of the plants increased with an ECa level of up to 3.0 dS m^{-1} .

Saline irrigation water was prepared by adding the salts NaCl, $CaCl_2.2H_2O$ and $MgCl_2.6H_2O$ in an equivalent ratio of 7:2:1, which is the predominant ratio in the main sources of water available for irrigation in the north-east of Brazil (MEDEIROS, 1992), following the relationship between ECa and salt concentration (mmolc L^{-1} = EC x 10) (RHOADES et al. 1992). The doses of N were determined based on the standard average dose recommended by Novais et al. (1991). With 100 mg dm^{-3} of soil corresponding to the 100% dose.

The experiment used the Morada Nova soursop cultivar, which, according to São José (2014), is a genetic material favoured by farmers in the northeast, as well as being the most widely used by seedling nurseries. The seeds were obtained from fruit harvested from a commercial orchard (Fazenda Boi Bravo) located in the municipality of Sousa - PB. The seeds were extracted by hand and then air-dried to break dormancy. This process consisted of immersing the seeds in a solution of gibberellic acid at a concentration of 750 mg L^{-1} , for a period of 9 hours (SANTOS et al., 2015).

2.4 Seedling production

The experiment was conducted using bags with a capacity of 1.2 dm^3 and holes in the sides to allow free drainage. These were placed on metal benches at a height of 0.8 metres from the ground (figure 1).

Figure 1: Soursop seedlings on a metal bench

The bags were filled with a substrate made up of soil + sand + bovine manure (well tanned) in proportions (by volume) of 82, 15 and 3% respectively. The organic matter was added in such a way as not to mask the effects of the treatments, but only to improve the soil's physical, chemical and biological characteristics. The soil's physical and chemical characteristics (Table 1) were obtained according to Claessen (1997) and analysed at the Soil and Plant Nutrition Laboratory at CCTA/UFCG.

Table 1. Physical and chemical characteristics of the substrate used in the experiment

Textural classification	Apparent density $g\ cm^{-3}$	Total porosity %	Organic matter $g\ kg^{-1}$	P $mg\ dm^{-3}$	Assortative complex			
					Ca^{2+}	Mg^{2+}	In^+	K^+
					-------- cmolc dm^{-3} --------			
Sandy loam	1,38	47,00	32	17	5,4	4,1	2,21	0,28

Saturation extract										
pHes	CEes $dS\ m^{-1}$	Ca^{2+}	Mg^{2+}	K^+	In^+	Cl^-	SO_4^{2-}	CO_3^{2-}	$HCO_3^{,}$	Saturation %
		------------------------- mmolc dm^{-3} -------------------------								
7,41	1,21	2,50	3,75	4,74	3,02	7,50	3,10	0,00	5,63	27,00

On 13 December 2015, two seeds were sown per bag at a depth of 1.5 cm. Emergence began at 20 days after sowing (DAS) and continued until the 40th day. At 45 DAS, thinning was carried out, leaving only the most vigorous seedling per bag, which was taken to the seedling stage.

During the germination and seedling emergence period, the substrate was kept at humidity close to field capacity, using local water supply (ECa of 0.3 dS m-1). For phytosanitary control, sprays were carried out using Organophosphate insecticide at a concentration of 150 ml 100 L^{-1} to control whitefly, according to the manufacturer's recommendations. Spraying was carried out at 5pm.

2.5 Application of treatments

The application of saline levels began at 7 DAE with daily manual irrigations, according to the treatments. Irrigations were based on the plant's water needs, using the drainage lysimetry process (20 bags with a collector in each), with daily application of the volume retained in the bags, determined by the difference between the volume applied and the volume drained in the previous irrigation (BERNARDO et al., 2006), carried out twice a day, in the early morning and late afternoon. A leaching fraction of 0.15 was also applied every fortnight, based on the volume applied during this period, in order to reduce the salinity of the substrate's saturation extract.

The nitrogen fertiliser was applied in 13 instalments, weekly from seven days after the plants had fully emerged, using Urea (45% N) as the source. Applications were made manually via irrigation water with an electrical conductivity of 0.3 dS m^{-1} for all treatments. The plants were managed during the seedling phase, for a period of 100 days (after germination), which is adequate time for transplanting as recommended by

Andrade et al. (2014).

2.6 Characteristics analysed

At 45 and 90 DAT, the growth characteristics of plant height (AP), stem diameter (DC), number of leaves (NF) and leaf area (AF) were assessed, as well as specific leaf area (AFE) and leaf area ratio (RAF). The absolute (TCAdc) and relative (TCRdc) growth rates of stem diameter were analysed between 15-45 and 15-90 DAT.

The AP (Figure 2A) was measured from the soil surface to the point of insertion of the apical meristem (in cm). The DC (Figure 2B) was measured at a height of 3 cm above ground level using a digital caliper (mm). When determining the NF, we considered the leaves whose leaf limb was fully open.

AF (Figure 2C) was determined according to the recommendations of Almeida et al. (2006), using equation 1, where AF = leaf area (dm^2) and X= is the product of length and width (cm).

$$AF = 5,71 + 0,647X \qquad R^2 = 0,91 \qquad \text{eq. 1}$$

Figure 2: Measurement of plant height (A), stem diameter (B) and leaf area (C).

The TCAdc is used to get an idea of the average speed of growth over the observation period, while the

TCRdc considers growth in relation to what the plant had previously (pre-existing material), i.e. it is a measure of how quickly a plant grows when compared to its initial size. These were determined according to the methodology described by Benincasa (2003), as shown in Equations 2 and 3.

$$TCAdc = \frac{(DC_2 - DC_1)}{(t_2 - t_1)} \quad eq.2 \qquad TCRdc = \frac{(lnDC_2 - lnDC_1)}{(t_2 - t_1)} \quad eq.3$$

In which:

TCAdc = absolute growth rate of stem diameter (mm day)[,-1]

TCRdc = Relative stem diameter growth rate (mm mm[-1] day)[,-1]

DC1 = stem diameter (mm) at time t_1,

DC2 = stem diameter (mm) at time t_2, ln = natural logarithm.

The RAF and AFE variables were determined according to eq. 4 and 5 (BENINCASA, 2003).

$$RAF = \frac{AF}{MSPA} \quad eq.4 \qquad AFE = \frac{AF}{MSF} \quad eq.5$$

In which:

AF: Leaf area (dm)[2]

MSPA: Dry phytomass of the aerial part (g)

MSF: Leaf dry phytomass (g)

2.7 Statistical analysis

The variables were assessed by analysing variance using the F test (P < 0.05 probability) and, in cases of significant effect, linear and quadratic polynomial regression analysis was carried out using the SISVAR statistical software (FERREIRA, 2011). The regression was chosen according to the best fit based on the coefficient of determination (R^2) and taking into account a probable biological explanation.

3 RESULTS AND DISCUSSION

Based on the summaries of the analyses of variance, there was a significant effect (p<0.01) of the salinity levels of the irrigation water on plant height (PH) and stem diameter (SD) at 45 and 90 DAT, and on leaf area (LA) and number of leaves (NF) of the soursop seedlings at 90 DAT (Table 2).

Despite the variation in the doses of nitrogen fertiliser (30 mg of N dm[-3]) given to the plants, there was no significant effect (Table 2) on the variables studied, indicating that the salinity of the water may have compromised the absorption of N due to ionic competition at the adsorption sites (NOBRE et al., 2010). Bosco et al. (2009) explain that plants grown under salinity tend to absorb less nitrogen while the levels of Cl absorbed and accumulated are increased.

Table 2. Summaries of the analyses of variance for plant height (PH), stem diameter (SD), leaf area (LA) and number of leaves (NF) of soursop seedlings irrigated with water of different salinities and nitrogen fertilisation at 45 and 90 days after treatment (DAT).

Source of variation	GL	MEAN SQUARES							
		AP		DC		AF		NF	
		45	90	45	90	45	90	45	90
Salt (dS m)$^{-1}$	4	$28,51^{**}$	$108,79^{**}$	$1,02^{**}$	$2,62^{**}$	$1,07^{ns}$	$8,64^{**}$	$4,23^{ns}$	$34,29^{**}$
Linear Regulation	1	$24,41^{ns}$	$146,30^{*}$	$0,01^{ns}$	$0,82^{ns}$	$2,06^{ns}$	$25,77^{**}$	$6,07^{ns}$	$90,13^{**}$
Quadratic rule	1	$82,57^{**}$	$230,04^{**}$	$3,17^{**}$	$8,61^{**}$	$1,51^{ns}$	$3,14^{ns}$	$6,48^{ns}$	$40,02^{**}$
Doses of N (%)	3	$2,96^{ns}$	$58,33^{ns}$	$0,13^{ns}$	$0,74^{ns}$	$0,86^{ns}$	$1,79^{ns}$	$3,77^{ns}$	$1,00^{ns}$
Linear Regulation	1	$2,13^{ns}$	$3,37^{ns}$	$0,39^{ns}$	$2,18^{*}$	$0,27^{ns}$	$1,18^{ns}$	$1,61^{ns}$	$2,61^{ns}$
Quadratic rule	1	$2,19^{ns}$	$30,31^{ns}$	$2x10^{-3ns}$	$0,02^{ns}$	$1,32^{ns}$	$0,61^{ns}$	$3,26^{ns}$	$0,26^{ns}$
Int. (S x N)	12	$6,83^{ns}$	$23,63^{ns}$	$0,04^{ns}$	$0,36^{ns}$	$0,89^{ns}$	$4,01^{ns}$	$7,38^{ns}$	$8,29^{ns}$
Block	3	$427,32^{**}$	$658,87^{**}$	$9,08^{**}$	$6,11^{**}$	$2,76^{*}$	$2,37^{ns}$	$2,21^{ns}$	$0,95^{ns}$
CV (%)		11,65	12,83	8,83	8,88	20,94	19,86	14,36	10,30

ns, **, * respectively not significant, significant at $p < 0.01$ and $p < 0.05$.

According to the regression equations (Figure 3A), it can be seen that increasing the salinity of the irrigation water had a quadratic effect (p<0.01) on plant height at 45 and 90 DAT, with an increase in PA up to the ECa level of 1.6 dS m^{-1}

(34.68 cm) and 1.5 dS m^{-1} (44.09 cm), respectively, and from this electrical conductivity onwards there were reductions. Therefore, given the reductions in PA, the sensitivity of the soursop seedlings to the increase in water salinity can be seen. It appears that the water deficiency, induced by the osmotic effect, may have caused morphological and anatomical changes in the plants to the point of impairing water absorption and the transpiration rate (SILVA et al., 2008).

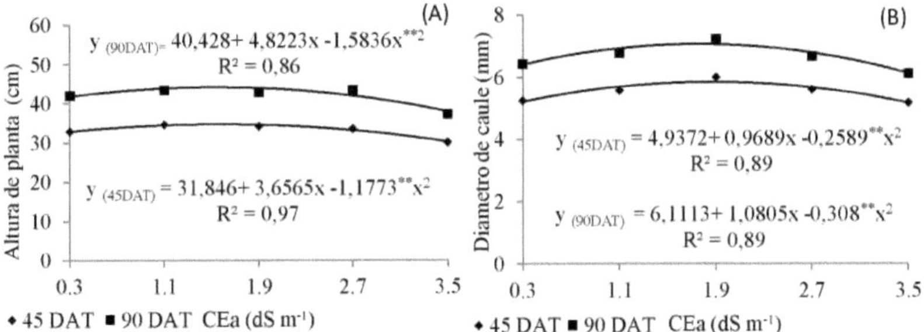

Figura 3. Plant height (AP) (A), stem diameter (DC) (B) at 45 and 90 days after application of the treatments - DAT, and of soursop seedlings as a function of irrigation water salinity - ECa.

The increase in the electrical conductivity of the irrigation water led to a reduction (p<0.01) in stem diameter at 45 and 90 DAT, and according to the regression equations (Figure 3B), there was a quadratic behaviour, where the highest estimated DC values of 5.84 and 7.05 mm were obtained at ECa of 1.9 and 1.6 dS m⁻¹ respectively. It can be seen that, initially, the soursop had greater tolerance to stress, i.e. in the period up to 45 DAT. According to Cavalcante et al. (2001), this superiority does not indicate that soursop seedlings can be produced with water with a saline concentration equal to or greater than 1.9 dS m⁻¹, since successive irrigations can increase the saline character of the substrate to the point where it reaches values not tolerated by the crop and, as a result, increase the loss of seedling quality. This was observed at 90 DAT, when tolerance was reduced to ECa = 1.6 dS m⁻¹. This reduction in diameter may possibly be a response to the closure of stomata and a reduction in gas exchange, consequently reducing the absorption of water and nutrients by the plants, which results in lower growth (LIMA et al., 2015).

Leaf area was only affected by salt stress at 90 DAT, with a linear decrease in PA of around 7.23%, i.e. 0.058 dm² per unit increase in ECa, i.e. when plants were subjected to an ECa of 3.5 dS m⁻¹ they had a 1.85 dm2 reduction in stem diameter compared to those under 0.3 dS m⁻¹ (Figure 4A). As the time of exposure to salts increases, plants are exposed to a greater degree of water retention, reaching a point where they don't have enough energy to absorb water in the proportion necessary to carry out their metabolic activities (NOBRE et al., 2014). Therefore, under conditions of water deficiency induced by the osmotic effect, it is common for morphological changes to occur in plants, such as smaller and fewer leaves, which also contributes to lower water consumption by the plant (COELHO et al., 2013).

According to Figure 4B, for the number of leaves (NF) at 90 DAT, there was a better fit of the data in quadratic regression by increasing the ECa, with the highest value corresponding to 18.86, which was obtained when the plants were irrigated with ECa water of 1.2 dS m⁻¹. The reduction in the number of leaves may be the result of the plant's adaptation mechanisms to salt stress, reducing the transpiring surface. In this way, the reduction in the number of leaves under such conditions is relevant to maintaining a high water potential in the plant (NOBRE et al., 2014).

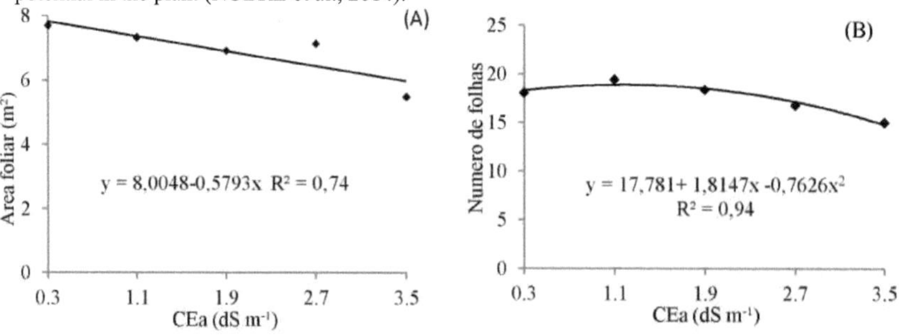

Figura 4.   Leaf area (A) and number of leaves (B) of soursop as a function of the electrical conductivity of the irrigation water, 90 days after application of the treatments - DAT.

Table 3 shows a significant effect (p<0.01) of the water salinity factor on the absolute diameter growth rate at 15-45 and 15-90 DAT, for the relative growth rate in the interval between 15-90 DAT and on the specific leaf area and leaf area ratio at 45 DAT. With regard to the nitrogen dose factor, as well as the interaction between the factors (water salinity x N doses), there was no significant effect (p>0.05) for any of the variables studied.

Table 3. Summaries of the analyses of variance for absolute (TCAdc) and relative (TCRdc) growth rate of stem diameter at 15-45 DAT and 15-90 DAT, specific leaf area (AFE) and leaf area ratio (RAF) of soursop seedlings assessed at 45 and 90 DAT, with the seedlings irrigated with water of different salinities and nitrogen fertilisation.

Source of variation	GL	MIDDLE SQUARE							
		TCAdc		TCRdc		AFE		RAF	
		15-45	15-90	15-45	15-90	45	90	45	90
Salt (dS m)$^{-1}$	4	$3x10^{*-4}$	$2x10^{-4}$**	$9x10\text{-}6^{ns}$	$5x10^{'6*}$	$0,49*$	$3,53^{ns}$	$0,24$**	$0,77^{ns}$
Linear Regulation	1	$2x11^{-4ns}$	$3x10^{-4*}$	$1x10^{-5ns}$	$1x10^{-5*}$	$1,44$**	$5,03^{ns}$	$0,61$**	$0,90^{ns}$
Quadratic rule	1	$6x10^{-4*}$	$6x10\text{-}4$**	$1x10^{-5ns}$	$7x10^{-6*}$	$0,18^{ns}$	$0,39^{ns}$	$0,30$**	$0,45^{ns}$
Doses of N (%)	3	$_1X10_{-4\,ns}$	$9X\text{-}5ns_{10}$	$5x10^{-6ns}$	$_2X\text{-}6ns_{10}$	$0,01^{ns}$	$2,38^{ns}$	$0,01^{ns}$	$0,45^{ns}$
Linear Regulation	1	$_1X10_{-4\,ns}$	$2x10^{-4}$	$8x10^{-6\,ns}$	$6x10^{-6ns}$	$8x10^{-4ns}$	$0,06^{ns}$	$0,01^{ns}$	$0,90^{ns}$
Quadratic rule	1	$_2X10_{-6\,ns}$	$1x10^{-5ns}$	$1x10^{-6\,ns}$	$1x10^{-6ns}$	$0,01^{ns}$	$3,78^{ns}$	$_1X10_{-4\,ns}$	$0,45^{ns}$
Int. (S x N)	12	$_6X\text{-}5ns_{10}$	$_7X\text{-}5ns_{10}$	$_4X10_{-6\,ns}$	$_2X\text{-}sns_{10}$	$0,16^{ns}$	$2,28^{ns}$	$0,03^{ns}$	$0,57^{ns}$
Block	3	$4x10^{-4*}$	$3x10^{-5ns}$	$9x10^{-6ns}$	$1x10^{-4}$**	$0,263^{ns}$	$4,43^{ns}$	$0,15*$	$1,08^{ns}$
CV (%)		34,98	20,44	18,96	19,72	30,52	64,78	25,88	52,22

ns, **, * respectively not significant, significant at p < 0.01 and p < 0.05;

The TCAdc was influenced by the increase in salinity of the irrigation water and according to the regression equations (Figure 5A) there was a quadratic effect for both evaluation intervals (15-45 and 15-90 DAT). In the period corresponding to the 15-45 interval, the highest TCAdc value (0.033 mm day^{-1}), as well as for the stem diameter variable, was observed when the seedlings were subjected to an electrical conductivity of 1.9 dS m·i; in the 15-90 interval, the highest value (0.039 mm day^{-1}) was seen at an ECa of 2.0 dSm· i. It is understood that the reduction in growth from these points onwards (1.9 and 2.0 dS m· 1, respectively) may be related to energy expenditure for the synthesis of osmotically active organic compounds necessary for compartmentalisation processes in the regulation of ion transport (LOPES & KLAR 2009).

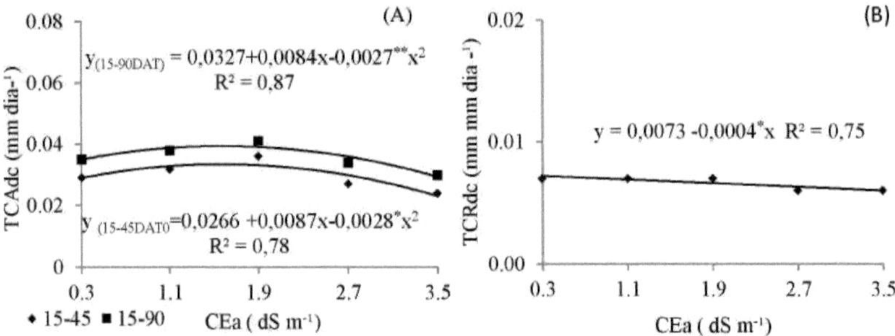

Figura 5. Absolute growth rate of soursop seedlings (TCAdc) in the period between 15-45 and 15-90 DAT (A), Relative growth rate of stem diameter (TCRdc) in the period 15-90 DAT (B) as a function of water salinity - ECa.

From the regression equation (Figure 5B), it can be seen that the increase in salinity in the irrigation water caused a decreasing linear effect on the TCRdc in the 1590 DAT period, where the plants irrigated with EC water of 3.5 dS m⁻¹ had a reduction in TCRdc of 0.007 mm mm⁻¹ day⁻¹ (17.53%) in relation to the plants irrigated with water of 0.3 dS m⁻¹ . The decrease in relative growth may be due to the growing accumulation of salts in the soil through irrigation over the course of the experimental period, which contributed negatively to water absorption by the plants due to the osmotic effect and/or the toxic effect of salts on the plants in such a way as to reduce plant growth (TRAVASSOS et al., 2012), as there was a lack of uniformity in the stand, burning of the edges and leaf colour not characteristic of the crop.

The specific leaf area was significantly influenced at 45 DAT by the salinity of the irrigation water, and according to the regression equation (Figure 4 A) there was an increasing linear response, in the order of 13.65% per unit increase in ECa, i.e, the highest EFA (1.49 cm² g⁻¹) was obtained at a salt level of 3.5 dS m⁻¹ , i.e. there was an increase in EFA of 43.69% when the plants were subjected to the highest electrical conductivity (3.5 dS m⁻¹) in relation to the plants irrigated with water with an ECa of 0.3 dS m⁻¹ . In this context, there is a greater tolerance of soursop seedlings to irrigation water salinity in the initial growth phase (45 DAT), as they achieved satisfactory growth even when subjected to an ECa of 3.5 dS m⁻¹ .

According to Tester & Davenport (2003), the response of plants to salinity can vary depending on the species, environmental factors, type and salinity level of the water and/or soil, so when plants are exposed to stress they tend to develop tolerance mechanisms such as: intracellular compartmentalisation and biochemical signalling; and control of the absorption and internal transport of salts, as well as the accumulation of Na. In the present study, it is likely that the soursop seedlings developed some kind of tolerance mechanism, especially in the initial growth phase, in order to contribute to obtaining greater EFA.

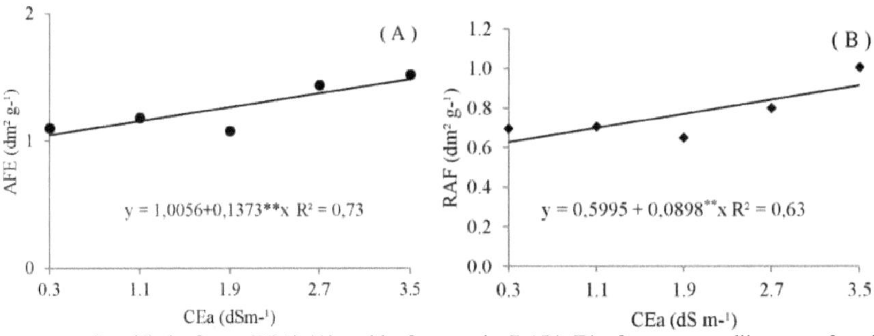

Figure 6 - Specific leaf area (SFA) (A) and leaf area ratio (LAR) (B) of soursop seedlings as a function of irrigation water salinity at 45 days after treatment (DAT).

As with EFA, it can be seen (Table 3) that RAF was influenced ($p<0.01$) by the increase in salinity of the irrigation water only at 45 DAT, and according to the regression equation (Figure 6B) there was an increasing linear behaviour, with an increase in RAF of 14.97% per unit increase in ECA, i.e. an increase of 0.28 dm² (52.89%) in plants under ECA of 3.5 dS m⁻¹ compared to plants irrigated with supply water. It can be seen that at 45 DAT, the plants under higher levels of water salinity had a greater amount of photosynthesising material in relation to the dry mass of the aerial part. This may be related to the adjustment of the crop's water potential, allowing it to absorb water under these conditions (COELHO et al. 2014).

4 CONCLUSIONS

The growth of Morada Nova soursop seedlings subjected to different levels of water salinity was less affected in the initial phase (45 DAT).

In the production of soursop seedlings, water with an ECa of 2.1 dS m⁻¹ can be used, as it provides an acceptable average reduction in growth of 10 per cent.

Doses of N higher than 70 mg N dm⁻³ of soil do not attenuate salt stress or promote greater growth in Morada Nova soursop seedlings.

The interaction between nitrogen doses and water salinity levels did not affect the growth phase of Morada Nova soursop seedlings.

5 REFERENCES

ALMEIDA, G.; SANTOS, J.; ZUCOLOTO, M.; VICENTINI, V.; MORAES, W., BREGONCIO, I.; COELHO, R.. Estimation of leaf area of graviola *(Annona muricata* L.) using linear dimensions of the leaf limb. **Revista UNIVAP**, v. 1, n.24, p.1035-1037, 2006.

ANDRADE, B. B. de, MELO, B. DE, SILVA, A. A. da; SOUZA, C. H. E. de. Containers and proportions of chicken litter in the production of soursop seedlings. **Revista Verde de agroecologia e desenvolvimento sustentável**, v. 9, n. 5, 116 - 123, 2014.

BERNARDO, S.; SOARES, A. A.; MANTOVANI, E. C. **Manual de irrigação**. 8. ed. Viçosa: UFV, p. 625, 2006.

CAVALCANTE, L. F.; CARVALHO, S. D.; LIMA, E. D.; FEITOSA FILHO, J. C.; SILVA, D. A. Initial development of soursop under sources and levels of water salinity. **Revista Brasileira de Fruticultura**, v. 23, n.2, p.455-459, 2001.

CAVALCANTE, L. F; VIEIRA, M. S.; SANTOS, A. F.; OLIVEIRA, W. M.; NASCIMENTO, J. A. M. Saline water and liquid bovine manure in the formation of Paluma guava seedlings. **Revista Brasileira de Fruticultura**, v. 32, n.1, p. 251261, 2010.

CLAESSEN, M. E. C. (org.). **Manual of soil analysis methods**. 2. ed. rev. atual. Rio de Janeiro: Embrapa-CNPS, (Embrapa-CNPS. Documentos, 1), p. 212, 1997.

COELHO, M. A.; SONCIN, N. B. **Geografia do Brasil**. São Paulo: Moderna, 368 p., 1982.

COELHO, B.; BARROS, M. F. C.; BEZERRA NETO, E.; CORREA, M. M. Water behaviour and growth of vigna beans grown in salinised soils. **Brazilian Journal of Agricultural and Environmental Engineering**. v.17, n. 4,p. 379-385, 2013.

COELHO, D. S. et al. Germination and initial growth of forage sorghum varieties subjected to salt stress. **Brazilian Journal of Agricultural and Environmental Engineering**. v.18, n.1, p.25-30, 2014.

COSTA, A. M. G.; COSTA, J. T. A.; JUNIOR, A. T. C.; CORREIA, D.; MEDEIROS FILHO, S. Influence of different substrate combinations on the formation of soursop rootstocks (Annona muricata L.). **Revista Ciência Agronomica**, v.36, n.3, p. 299-305. 2008.

FERREIRA, D. F. Sisvar: a computer system for statistical analysis. **Ciência e agrotecnologia**, v. 35, n. 6, p. 1039-1042, 2011.

HOLANDA FILHO, R. S.; SANTOS, D. B. dos; AZEVEDO, C. A. de, COELHO, E. F.; LIMA, V. L. de. Saline water on soil chemical attributes and nutritional status of cassava. **Revista Brasileira de Engenharia**

Agrícola e Ambiental, v.15, n.1, p. 60-66, 2011.

LIMA, G. S. de; NOBRE, R. G.; GHEYI, H. R.; ANJOS SOARES, L. A. dos; SILVA, S. S.. Morphophysiological responses of papaya as a function of irrigation water salinity and nitrogen fertilisation. **Irriga**, v. 19, n. 1, p. 130-136, 2014.

LIMA, L. A.; OLIVEIRA, F. A. de; ALVES. R DE C.; LINHARES. P. S. F.; MEDEIROS. A. M. A. de; BEZERRA. F. M. S. Tolerance of aubergine to salinity in irrigation water. **Revista Agroambiente**, v.9, n.1, p.27-34, 2015.

LOPES, T. C.; KLAR, A. E. Influence of different salinity levels on morphophysiological aspects of Eucalyptus urograndis seedlings. **Irriga**, v.14, n.1, p.68-75, 2009.

MEDEIROS, J. F. Quality of irrigation water and evolution of salinity on properties assisted by "GAT" in the states of RN, PB and CE. Campina Grande: Federal University of Paraíba, 1992. 173 f. Master's thesis.

NASCIMENTO, H. H. C. do; SANTOS, C. A. dos.; FREIRE, C. S.; SILVA, M. A. da.; NOGUEIRA, R. J. M. C. Osmotic adjustment in jatobá seedlings subjected to salinity in hydroponic media, **Revista Árvore**, v. 39, n. 4, 2015.

NEVES, A. L. R.; LACERDA, C. F.; GUIMARÃES, F. V. A.; HERNANDEZ, F. F. F.; SILVA, F. B.; PRISCO, J. T.; GHEYI, H. R. Biomass accumulation and nutrient extraction by string bean plants irrigated with saline water at different stages of development. **Ciência** Rural, v.39, n.3, 758-765, 2009.

NOBRE, R. G.; GHEYI, H. R.; CORREIA, K. G.; SOARES, F. A. L.; ANDRADE, L. D. Sunflower growth and flowering under salt stress and nitrogen fertilisation. **Revista Ciência Agronómica**, v. 41, n.3, p. 358-365, 2010.

NOBRE, R. G.; LIMA, G. S. de; GHEYI, H. R., SOARES, L. A. A. dos; SILVA, A. O. da. Growth, consumption and efficiency of water use by papaya under saline and nitrogen stress. **Revista Caatinga**, v.27, n.2, p.148-158, 2014.

NOVAIS, R. D.; NEVES, J. C. L.; BARROS, N. D.; OLIVEIRA, A. D.; GARRIDO, W. E.; ARAUJO, J. D.; LOURENÇO, S. Trials in a controlled environment. In: OLIVEIRA A. J. (ed). **Research methods in soil fertility**. Brasília-DF: Embrapa - SEA, 1991, p. 189-253.

OLIVEIRA NETO, E. A. de.; SANTOS, D. C. da.; SANTOS, Y. M. G. dos, Agroindustrial utilisation of soursop (Annona muricata L.) for production of liqueurs: Sensory evaluation, **Journal of Biotechnology and Biodiversity.** v. 5, n.1, p. 33-42, 2014.

RHOADES, J. P.; KANDIAH, A.; MASHALI, A. M. **The use saline waters for crop production** (Org). Rome: FAO, 1992.

SANTOS, H. H. D.; MATOS, V. P.; ALBUQUERQUE, A. P. C. da ; SENA, L. H. MOURA de; FERREIRA, E. G. B. S. de. Morphology of fruits, seeds and seedlings of Averrhoa bilimbi L. from two stages of maturation. **Ciencia Rural,** v. 44 ,n.11, 1995-2002, 2014.

SANTOS, T. H. ; SILVA, G. V. ; SANTOS, A. S. ; NOBRE, R. G. . Evaluation of methods for overcoming pine cone (Annona Squamosa L.) dormancy in the Paraíba hinterland. In: XIX Brazilian Seed Congress, 2015, Foz de Iquaçu - PR. XIX Brazilian Seed Congress. Foz de Iquaçu - PR, 2015.

SÃO JOSÉ, A. R.; PIRES, M. M. de; FREITAS, A. L. G. E. de; RIBEIRO, D. P.; PEREZ, L. A. A. Current developments and prospects for Anonaceae in the world. **Revista Brasileira de Fruticultura,** v. 1, n.36, p. 86-93, 2014.

SILVA, S. M.; ALVES, A. N.; GHEYI, H. R.; BELTRÃO, N. D. M.; SEVERINO, L. S.; SOARES, F. A. Development and production of two cultivars of papaya under salt stress, **Revista Brasileira de Engenharia Agrícola e Ambiental,** v.12, n.4, p. 335-342, 2008.

SOUSA, A. B. O.; BEZERRA, M. A.; FARIAS, F. C. Germination and initial development of common cashew clones under irrigation with saline water. **Revista Brasileira de Engenharia Agrícola e Ambiental,** v.15, n.4, p. 390-394, 2011.

TESTER, M.; DAVENPORT, R. Na+ tolerance and Na+ transport in higher plants. **Annals of Botany,** v. 91, p. 503-527, 2003.

TRAVASSOS, K. D.; GHEYI, H. R.; SOARES, F. A. L.; BARROS, H. M. M., SILVA DIAS, N. da, UYEDA, C. A.; SILVA, F. V. da. Growth and development of sunflower varieties irrigated with saline water. **Irriga,** v.1, n.1, p. 324 - 339, 2012.

CHAPTER 3

PHYTOMASS AND QUALITY OF SOURSOP SEEDLINGS WITH DIFFERENT SALINE WATERS AND NITROGEN DOSES

SUMMARY

The use of saline water in agricultural activities should be considered an important alternative for rational exploitation, especially in regions where water scarcity often occurs, such as the Brazilian semi-arid region. The aim of this research was to assess the phytomass and quality of 'Morada Nova' soursop seedlings irrigated with water of different salinities and fertilised with doses of nitrogen. The experiment was carried out in a greenhouse at the Centre for Agri-Food Sciences and Technology of the Federal University of Campina Grande (Pombal - PB), using a randomised block experimental design in a 5 x 4 factorial scheme, corresponding to five levels of electrical conductivity of the water (0.3; 1.1; 1.9; 2.7 and 3.5 dS m^{-1}) and four doses of N (70, 100, 130 and 160 mg of N dm^{-3} of soil), with four replicates and two plants per plot. At 90 days after the treatments were applied, the fresh and dry phytomass of the stem and leaves and of the aerial part, roots and total plant were assessed, as well as the Dickson seedling quality index. Irrigation with water with an electrical conductivity of up to 2.0 dS m^{-1} enables the production of Morada Nova soursop seedlings with an acceptable reduction in growth. The dose of 70 mg dm^{-3} of nitrogen stimulates phytomass production and seedling quality. There was no interaction between the factors on the variables assessed in the seedlings 90 days after the treatments were applied.

Keywords: Annona muricata L., salt stress, seedling production.

1 INTRODUCTION

Belonging to the anonaceae family and considered the most tropical among them, the soursop tree *(Annona muricata,* L.) occupies a promising position in Brazilian fruit growing, especially in the Northeast of Brazil, where its consumption has increased, whether in *natura* or industrially processed, due to its nutritional importance and the ways in which it can be used in human nutrition, as well as the medicinal properties of its leaves, fruit, seeds and roots (CAMPOS et al., 2008).

Despite the importance of this fruit tree for northeastern Brazil, rainfall in the semi-arid area is irregular and the water resources available for irrigation are often scarce and vary in spatial and temporal distribution. As a result, the quality of these water resources also varies, and water sources with a high concentration of salts are common (BEZERRA et al. 2010).

According to Sairam & Tyagi (2004), it is estimated that 20 per cent of the world's cultivated land, which corresponds to around half of the irrigated areas, is affected by salts. Concentrations of salts in irrigation water and/or soil in quantities greater than those tolerated by plants can cause a reduction in the soil's osmotic potential and thus reduce the availability of water for plants, as well as promoting toxicity and nutritional imbalance (NOBRE et al., 2010). Thus, the use of saline water in irrigation is subject to crop tolerance to salinity, as well as irrigation and fertilisation management practices, which should avoid environmental impacts and consequent damage to crops (AMORIM et al., 2010; RIBEIRO et al., 2015).

In the production process, in addition to the importance of an adequate supply of water in terms of quantity and quality, fertilisation is a key factor in obtaining positive results. Among the macronutrients required by plants, nitrogen is one of the most important because it is part of the plant's structure, being a component of amino acids, proteins, enzymes, RNA, DNA, ATP, chlorophyll, among other molecules, and it is also a nutrient that is directly related to characteristics linked to plant growth (CHAVES et al., 2011). Furthermore, studies have shown that the accumulation of this organic solute increases the osmotic adjustment capacity of plants to salinity (SILVA et al., 2008).

The aim of this research was to assess the phytomass production and quality of Morada Nova soursop seedlings irrigated with different levels of water salinity and different doses of nitrogen fertiliser.

2 MATERIAL AND METHODS

2.1 Location of the experiment

The study was carried out in a greenhouse at the Federal University of Campina Grande, at the Centre for Agri-Food Sciences and Technology, Pombal-PB Campus, where the local geographic coordinates are "6°48'16" S, "37°49'15" O and an average altitude of 184m. According to the Koppen classification, adapted to Brazil, the region's climate is classified as BSh, hot semi-arid, with an average temperature of 28 °C, rainfall of around 750 mm per year^{-1} and average evaporation of 2000 mm (COELHO; SONCIN, 1982).

2.2 Experimental design and treatments

The randomised block experimental design was used in a 5 x 4 factorial scheme, with four replications and two plants per plot. The treatments consisted of a combination of five levels of electrical conductivity of the irrigation water - ECa (0.3; 1.1; 1.9; 2.7 and 3.5 dS m^{-1}) associated with four doses of nitrogen fertilisation (70; 100; 130 and 160% of the recommended dose for the crop, according to Novais et al. 1991), with the 100 per cent dose corresponding to 100 mg of N dm^{-3} of soil.

2.3 Treatment descriptions

The salt levels were selected on the basis of quotes by Cavalcante et al. (2001), who classified soursop in the initial growth phase as moderately sensitive to salinity, i.e. the biological yield of the plants increased with the ECa level of up to 3.0 dS m^{-1} .

The water with different saline levels for irrigation was obtained by adding different amounts of NaCl, CaCl2.2H2O and MgCl2.6H2O salts, in an equivalent ratio of 7:2:1, which is the predominant ratio in the main sources of water available for irrigation in the north-east of Brazil (MEDEIROS, 1992), following the relationship between ECa and salt concentration (mmolc L^{-1} = EC x 10) (RHOADES et al. 1992). The doses of N were determined based on the standard average dose recommended by Novais et al. (1991). With 100 mg dm-^{3} corresponding to the 100% dose.

The experiment used the Morada Nova soursop cultivar, which, according to São José (2014), is a genetic material favoured by farmers in the Northeast, as well as being the most widely used by seedling nurseries. We used seeds from ripe fruit harvested from a commercial orchard (Fazenda Boi Bravo) located in the municipality of Sousa - PB. The seeds were extracted by hand and then air-dried to break dormancy. This process consisted of immersing the seeds in a solution of gibberellic acid at a concentration of 750 mg L^{-1} , for a period of 9 hours (SANTOS et al.,2015).

2.4 Seedling production

Plastic bags with a capacity of 1.2 dm^3 were used to sow the seedlings. The sides of the bags had holes to allow free drainage. They were filled with a substrate made up of 82% soil, 15% sand and 3% tanned cattle manure. The addition of the manure was aimed at improving the soil's physical, chemical and biological properties in order to improve its water retention and infiltration capacity. The physical and chemical characteristics of the soil (Table 1) used in the experiment were obtained according to Claessen (1997) and analysed at the CCTA/UFCG Soil and Plant Nutrition Laboratory.

Table 4. Physical and chemical characteristics of the substrate used in the experiment

Textural classification	Apparent density g cm^{-3}	Total porosity %	Organic matter g kg^{-1}	P mg dm^{-3}	Assortative complex Ca^{2+}	Mg^{2+}	In$^+$	K$^+$
					-------- cmolc dm^{-3} ---------			
Sandy loam	1,38	47,00	32	17	5,4	4,1	2,21	0,28

pHes	CEes dS m^{-1}	Ca^{2+}	Mg^{2+}	K$^+$	In$^+$	Cl$^-$	SO$_4^{2-}$	CO$_3^{2-}$	HCO3$^-$	Saturation %
					Saturation extract ------------------------ mmolc dm^{-3} ------------------------					
7,41	1,21	2,50	3,75	4,74	3,02	7,50	3,10	0,00	5,63	27,00

The seeds were sown on 13 December 2015, with two seeds per bag at a depth of 1.5 cm. Seedling emergence began twenty days after sowing (DAS) and continued until the fortieth day. At 5 days after the seedlings had fully emerged, thinning was carried out, leaving only the most vigorous seedling. Before sowing, the soil was brought to field capacity and during the germination and emergence period, the seedlings were irrigated with local water supply (ECa = 0.3 dS m).$^{-1}$

2.5 Application of treatments

The application of the different salt levels began at 7 days after emergence (DAE), with daily irrigations carried out manually according to the treatments. Irrigation was based on the plants' water needs, determined by the drainage lysimeter process (using 20 bags with a collector in each), and was carried out twice a day, in the early morning and late afternoon at 5pm. Every fortnight, a leaching fraction of 0.15 was applied based on the volume applied during this period, in order to reduce the salinity of the substrate.

The application of nitrogen fertiliser also began at 7 DAE and was divided into 13 applications, carried out 7 days apart. The N source used was urea (45% N), and the applications were made via fertigation using water with an ECa of 0.3 dS m^{-1} for all treatments. The plants were grown during the seedling stage, i.e. for a period of 100 days (after germination), which is adequate time for transplanting, according to Andrade et al. (2014).

Cultivation during the growing season consisted of manually eliminating spontaneous plants, surface scarification of the soil and phytosanitary control using Organophosphate insecticide at a concentration of 150 mL 100 L^{-1} to control whitefly according to the manufacturer's recommendations.

2.6 Characteristics analysed

The fresh phytomass of the stems (FFC) and leaves (FFF) was determined, as well as the dry phytomass of the stems (FSC), leaves (FSF), aerial part (FSPA), root (FSR) and total (FST).

The stem of each plant was cut off close to the ground and the stem and leaves were separated and weighed immediately on a precision scale (0.001 g) to determine the FFC and FFF. After weighing the fresh mass, the different parts of the plant (leaves, stem and roots) were packed separately in duly labelled paper bags and dried in a forced air circulation oven at 65 °C until a constant mass was obtained, at which point the FSF, FSC and FSR were determined; the sum of the FSC and FSF was used to determine the FSPA, and the FSR was used to calculate the FST. The roots were extracted from the substrate using a 3 mm sieve and running water. The quality of the seedlings was determined using the Dickson quality index (DQI) for seedlings, using the formula of Dickson et al. (1960), described in equation 1.

$$IQD = \frac{(FST)}{(AP/DC) + (FSPA/FSR)} \qquad \text{Eq. 1}$$

In which:

IQD = Dickson quality index

AP = plant height (cm),

DC = stem diameter (mm).

FST = total plant dry phytomass (g)

FSPA = dry phytomass of the aerial part of the plant (g)

FSR = plant root dry phytomass (g)

2.   7Statistical analysis

Carried out in the same way as the previous chapter.

3 RESULTS AND DISCUSSION

As shown in Table 2, the salinity of the irrigation water had a significant effect on all the variables studied, i.e. fresh phytomass of leaf (FFF) and stem (FFC), dry phytomass of leaf (FSF), stem (FSC), root (FSR) and total (FST) and on the Dickson quality index (IQD) of soursop seedlings at 90 DAT. In contrast, nitrogen fertilisation and its interaction with water salinity did not significantly affect any of the variables analysed. This indicates that the different doses of nitrogen behaved in a similar way at different levels of salinity of the water used for irrigation.

Table 5 Summaries of the analyses of variance for fresh phytomass of leaf (FFF) and stem (FFC), dry phytomass of leaf (FSF), stem (FSC), root (FSR), aerial part (FSPA) and total plant (FST) and Dickson quality index (IQD) of soursop seedlings irrigated with water of different saline levels and under doses of nitrogen fertilisation at 90 DAT.

Sources of Variation	GL	MIDDLE SQUARE							
		FFF	FFC	FSF	FSC	FSR	FSPA	FMSt	IQD
Salt (dS m^{-1})	4	$74{,}99^{**}$	$22{,}65^{**}$	$6{,}76^{**}$	$2{,}19^{**}$	$1{,}03^{**}$	$15{,}25^{**}$	$23{,}82^{**}$	$0{,}27^{**}$
Linear Regulation	1	$188{,}85^{**}$	$24{,}36^{*}$	$18{,}26^{**}$	$3{,}42^{**}$	$1{,}47^{*}$	$38{,}34^{**}$	$53{,}88^{**}$	$0{,}40^{**}$
Quadratic Regulation	1	$105{,}44^{**}$	$54{,}03^{**}$	$8{,}33^{**}$	$5{,}29^{**}$	$2{,}55^{**}$	$22{,}27^{**}$	$41{,}16^{**}$	$0{,}60^{**}$
Doses of N (%)	3	$16{,}71^{ns}$	$6{,}77^{ns}$	$1{,}11^{ns}$	$0{,}21^{ns}$	$0{,}07^{ns}$	$1{,}91^{ns}$	$4{,}00^{ns}$	$0{,}01^{ns}$
Linear Regulation Quadratic	1	$0{,}24^{ns}$	$0{,}17^{ns}$	$0{,}01^{ns}$	$0{,}04^{ns}$	$0{,}06^{ns}$	$0{,}06^{ns}$	$0{,}20^{ns}$	$0{,}01^{ns}$
Regulation	1	$0{,}224^{ns}$	$0{,}60^{ns}$	$0{,}01^{ns}$	$0{,}02^{ns}$	$0{,}01^{ns}$	$0{,}26^{ns}$	$0{,}08^{ns}$	$0{,}00^{ns}$
Salt X Doses	12	$5{,}81^{ns}$	$3{,}32^{ns}$	$0{,}43^{ns}$	$0{,}16^{ns}$	$0{,}26^{ns}$	$1{,}40^{ns}$	$1{,}67^{ns}$	$0{,}04^{ns}$
Block	3	$33{,}92^{**}$	$43{,}08^{**}$	$3{,}27^{**}$	$5{,}16^{**}$	$2{,}08^{**}$	$1{,}29^{ns}$	$27{,}31^{**}$	$0{,}29^{**}$
Cv (%)		27,19	28,01	26,41	28,76	34,73	25,94	24,93	28,90

ns, **, * respectively not significant, significant at $p < 0.01$ and $p < 0.05$

The accumulation of fresh mass in the stems and leaves of soursop seedlings was significantly affected (p< 0.01) by the salinity of the irrigation water (Table 5).0.01) by the salinity of the irrigation water (Table 5), whose data was best fitted by quadratic equations (Figure 6A), obtaining greater fresh phytomass of stem and leaves when the seedlings were irrigated with water of 1.6 dS m^{-1} (8.23 g) and 1.2 dS m^{-1} (13.06 g), respectively, in which they obtained an increase of 1.15 g and 0.78 g compared to those irrigated with supply water.The reduction in FFF and FFC values from the saline levels described above may be attributed to the reduction in the osmotic potential of the soil solution, due to the excess salts present, making it difficult for plants to absorb water, causing them to expend greater energy in order to absorb water and nutrients, thus reducing their growth (NOBRE et al., 2010). Likewise, the effect of salts generally causes ionic toxicity to occur, and plants tend to close their stomata to reduce water loss through transpiration, resulting in a lower photosynthetic rate and, consequently, a reduction in the phytomass production of species under stress (CHEN & JIANG, 2010).

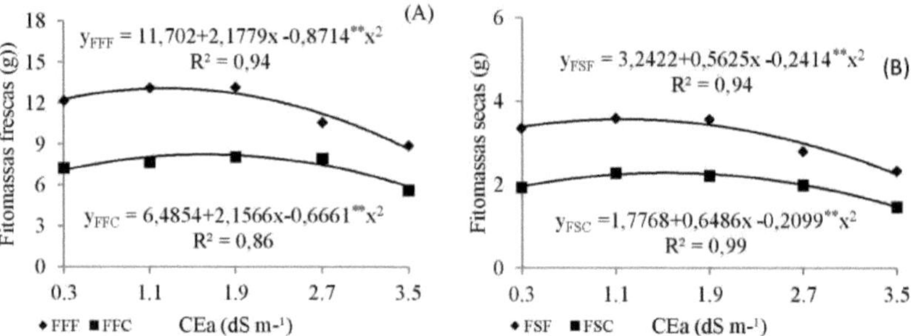

Figura 5. Fresh phytomass of stems - FFC and leaves - FFF (A) and dry phytomass of stems - FSC and leaves - FSF (B) of soursop seedlings as a function of irrigation water salinity at 90 days after treatment - DAT.

The salinity of the irrigation water had a significant influence (p<0.01) on the dry phytomass of the leaf and stem and according to the regression equations (Figure 6B), the model to which the data best fitted was quadratic, showing an increase in phytomass up to the levels of 1.2 dS m⁻¹ (3.56 g) for FSC and 1.5 dS m⁻¹ (2.27 g) for FSF. The reduction in dry phytomass from these ECa levels onwards is closely linked to the effects of the accumulation of soluble salts, which is a limiting factor for the development of most crops, and it can be deduced that this behaviour can be understood as a possible plant adjustment mechanism to reduce the effects of salinity, because plants can undergo morphological or physiological changes, such as a reduction in biomass, when subjected to saline stress (CENTENO et al., 2014). A similar finding was made by Távora et al. (2004), who found a reduction in the dry phytomass of young soursop plants when subjected to salt stress.

According to the regression equations (Figure 7A), it can be seen that the dry phytomass of the aerial part and the roots showed quadratic responses as a function of the increase in

⁻¹The maximum values of 6.42 (FSPA) and 1.54 g (FSR) were obtained when irrigating with water of 1.3 and 1.5 dS m⁻¹ , respectively, and when these were subjected to irrigation with ECa of 3,5 had a respective reduction of 2.80 g and 0.64 g in FSPA and FSR, so it can be seen from the regression functions that the greatest effect of salinity is evident in the aerial part of the plants, as seen by the greater decrease in FSPA compared to FSR. According to the behaviour of the phytomass of the aerial part and roots of soursop seedlings at 90 DAT (Figure 7A), the species was found to be tolerant up to an average water salinity level of 1.3 dS m⁻¹ , since up to this level the highest phytomass values were obtained. It should also be noted that as the ECa level increased, there was a reduction in phytomass production, since increasing the saline concentration in the soil solution reduces the soil's osmotic potential, making it difficult for plants to absorb water, promoting negative nutritional effects, toxicity and/or interfering with the availability of other ions (LIMA et al., 2014).

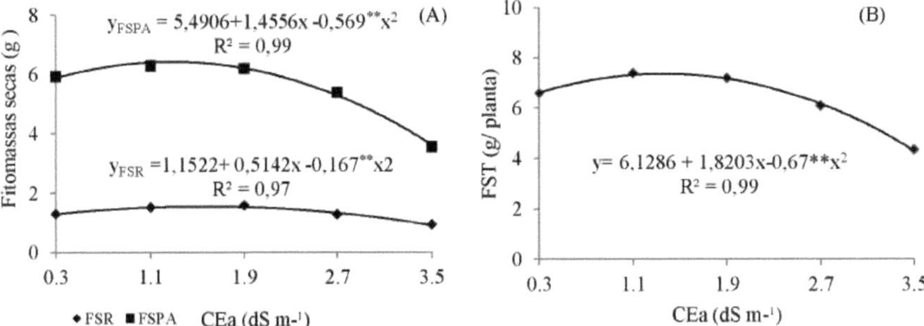

Figura 6. Root dry phytomass (RDS) and aerial part dry phytomass (APPS) (A) and total dry phytomass (TDS) (B) of soursop seedlings as a function of irrigation water salinity at 90 days after treatment (DAT).

The increasing salinity of the irrigation water also led to a quadratic behaviour in total dry phytomass (Figure 7B), with the highest FST value (7.36 g) obtained when the plants were irrigated with an ECa of 1.4 dS m⁻¹ , from which point there was a decrease in this variable. It should be emphasised that salt stress can cause a nutritional and physiological imbalance with a direct influence on the conversion of carbon assimilated by the plants, promoting reductions in the growth and accumulation of crop biomass (TAIZ & ZEIGER, 2009), which can be observed in this research through FST.

In the variable related to the quality of soursop seedlings (IQD), there was a significant effect (p<0.01) of the salinity levels of the irrigation water and according to Figure 8, the mathematical model that best fitted the data was the quadratic one, whose highest IQD (0.73) was obtained when the seedlings were irrigated with water with an ECa of 1.5 dS m⁻¹ , from which there was a reduction and the plants that were subjected to an ECa of 3.5 dS m⁻¹ showed an IQD of 0.41. This is interesting because even under conditions of saline stress, Morada Nova soursop seedlings had an IQD of more than 0.2, which is considered good quality seedlings for transplanting, according to the criteria established by Gomes (2001), since the higher the IQD value, the better the quality of the seedling.

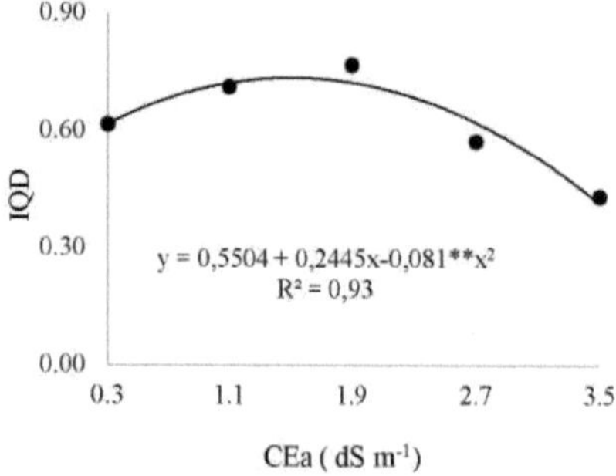

Figura 7. Dickson quality index (DQI) of hawthorn seedlings as a function of irrigation water salinity at 90 days after application of treatments - DAT.

4 CONCLUSION

Irrigation with water with an electrical conductivity of up to 2.0 dS m^{-1} enables the production of Morada Nova soursop seedlings with an acceptable reduction in growth.

The N dose of 70 mg dm^{-3} is sufficient for the production of phytomass and quality of soursop seedlings at 90 DAT.

There was no interaction between irrigation water salinity and nitrogen fertilisation on the variables assessed in Morada Nova soursop seedlings.

5 BIBLIOGRAPHICAL REFERENCES

ALMEIDA, G.; SANTOS, J.; ZUCOLOTO, M.; VICENTINI, V.; MORAES, W., BREGONCIO, I.; COELHO, R.. Estimation of leaf area of graviola *(Annona muricata* L.) using linear dimensions of the leaf limb. **Revista UNIVAP**, v. 1, n.24, p.1035-1037, 2006.

AMORIM, A. V.; GOMES FILHO, E.; BEZERRA, M. A.; PRISCO, J. T., LACERDA, C. F. Physiological responses of adult early dwarf cashew plants to salinity. **Revista Ciência Agronômica**, v. 41, n.1, p. 113-121, 2010.

ANDRADE, B. B. de, MELO, B. DE, SILVA, A. A. da; SOUZA, C. H. E. de. Containers and proportions of chicken litter in the production of soursop seedlings. **Revista Verde de agroecologia e desenvolvimento sustentável**, v. 9, n. 5, 116 - 123, 2014.

BERNARDO, S.; SOARES, A. A.; MANTOVANI, E. C. **Manual de irrigação.** 8. ed. Viçosa: UFV, p. 625, 2006.

BEZERRA, A. K. P.; LACERDA, C. F. de; HERNANDEZ, F. F. F.; SILVA, F. B. da; GHEYI, H. R. Cowpea/maize crop rotation using waters of different salinities. **Revista Ciência Rural,** v.40, n. 5, p.1075-1082, 2010.

CAMPOS, M. C. C.; MARQUES, F. J.; LIMA, A. G. de; MENDONÇA, R. M. N. de. Growth of soursop rootstock (Annona muricata, L.) in substrates containing increasing doses of kaolin tailings. **Revista de biologia e ciências da terra.** v. 8, n. 1, p. 61-66, 2008.

CENTENO, C. R. M.; SANTOS, J. B. dos; XAVIER, D. A.; AZEVEDO, C. A. V. de; GHEYIR, H. R. Production components of Embrapa 122-V2000 sunflower under water salinity and nitrogen fertilisation. **Revista Brasileira Engenharia Agrícola Ambiental,** v.18, (Suplemento), p. 39-45, 2014.

CHAVES, L. H. G.; GHEYI, H. R.; RIBEIRO, S. Water consumption and use efficiency for castor bean cultivar paraguaçu subjected to nitrogen fertilisation. **Engenharia Ambiental: pesquisa e tecnologia,** v. 8, n. 1, p-126-133, 2011.

CHEN, H. & JIANG, J. Osmotic adjustment and plant adaptation to environmental changes related to drought and salinity. **Environmental Reviews,** v.18, n.A, p.309-319, 2010.

CLAESSEM, M. E. C. (Obg). **Manual of soil analysis methods.** 2. ed. Rev. Atual. Rio de Janeiro: Embrapa-CNPS, p. 212, 1997 (Embrapa- CNPS. Documentos, 1).

DICKSON, A.; LEAF, A. L.; HOSNER, J. F. Quality appraisal of white spruce and white pine seedling stock in nurseries. **The Forest Chronicle,** v. 36, n. 1, p. 10-13, 1960.

FERREIRA, D. F. Sisvar: a computer system for statistical analysis. **Ciência e agrotecnologia,** v. 35, n. 6, p. 1039-1042, 2011.

GOMES, J. M. Parâmetros morfológicos na avaliação da qualidade de mudas de Eucalyptus grandis, produzidas em diferentes tamanhos de tubete e de dosagens de N-P- K., Universidade Federal de Viçosa, Viçosa, p. 112, 2001. Thesis (Doctorate in Forestry Sciences).

LIMA, G. S. de; NOBRE, R. G.; GHEYI, H. R.; SOARES, L. A. A. dos; SILVA, S. S. da. Morphophysiological responses of papaya as a function of irrigation water salinity and nitrogen fertilisation. **Irriga,** v. 19, n. 1, p. 130-136, 2014.

NOBRE, R. G.; LIMA, G. S. de; GHEYI, H. R., SOARES, L. A. A. dos; SILVA, A. O. Growth, consumption and efficiency of water use by papaya under saline and nitrogen stress. **Revista Caatinga**, v. 27, n. 2, p. 148-158, 2014.

NOVAIS, R. F.; NEVES J. C. L.; BARROS N. F. Trials in a controlled environment. In: OLIVEIRA A. J. (ed) **Research methods in soil fertility**. Brasília: Embrapa- SEA. p. 189-253. 1991.

RHOADES, J. P.; KANDIAH, A.; MASHALI, A. M. **The use saline waters for crop production**. V. 48. Rome: FAO, 1992.

RIBEIRO, P. H. P.; SILVA, S.; NETO, J. D.; OLIVEIRA, C. S. da; CHAVES, L. H. G. Growth and production components of sunflower as a function of irrigation with saline water and nitrogen fertilisation. **Engineering in Agriculture**, v.23 n.1, p.48-56, 2015.

SAIRAM, R. K.; TYAGI, A. Physiology and molecular biology of salinity stress tolerance in plants. **Current Science**, v.86, n.03, p.407-421, 2004.

SÃO JOSÉ, A. R.; PIRES, M. M. de; FREITAS, A. L. G. E. de; RIBEIRO, D. P.; PEREZ, L. A. A. Current developments and prospects for Anonaceae in the world. **Revista Brasileira de Fruticultura**, v. l, n.36, p. 86-93, 2014.

SILVA, E. C.; NOGUEIRA, R. J. M. C.; ARAÚJO, F. P.; MELO, N. F.; AZEVEDO NETO, A. D. Physiological responses to salt stress in young umbu plants. **Environmental and Experimental Botany**, v.63, p.147-157, 2008.

TAIZ, L.; ZEIGER, E. **Plant Physiology**. 4.ed. Porto Alegre: Artmed, 2009, p. 819.

TÁVORA, F. J. A. F.; LIMA, E. C. C.; HERNANDEZ, F. F. F. Mineral composition of roots, stems and leaves in young graviola plants subjected to salt stress. **Ciência. Agronómica**, Fortaleza, v. 35, N.1, p. 44 - 51, 2004.

Printed by Books on Demand GmbH, Norderstedt / Germany